GENESIS PROTOCOL

GENESIS PROTOCOL

Dawn of a New Era

A SCIENCE FICTION NOVEL BY

DOUGLAS D. BEATENHEAD

iUniverse, Inc.
Bloomington

GENESIS PROTOCOL
DAWN OF A NEW ERA

iUniverse books may be ordered through booksellers or by contacting:

iUniverse
1663 Liberty Drive
Bloomington, IN 47403
www.iuniverse.com
1-800-Authors (1-800-288-4677)

ISBN: 978-1-4759-2957-7 (sc)
ISBN: 978-1-4759-2958-4 (ebk)

Printed in the United States of America

iUniverse rev. date: 06/27/2012

CONTENTS

CHAPTER ONE

Well, this time it's three o'clock in the morning and I'm sweating and shaking all over again. It usually takes me an hour or so to recover from these horrific, nighttime phantasms. After waking up tonight, I did my usual bathroom routine, and then grabbed myself a beer. After four or five hours of sleep, I'm always awakened by a visit from a very strange and mystical man. We all know him-especially from our childhood. I've come to know him quite differently and in a tainted light, and he's become an absolute monster to me-a terrible and vile intrusion of my sleep: *Mr. Sandman*. My twilight life has been damned because of this man. He's persecuted me for over the last five years and I'm not sure how, or even if I can beat this thing. But if I don't soon, I'll find myself someday in a dreadful and gloomy room playing 'etch-o-sketch' and winning.

Tonight was the carcass dream. Many people know the stanch smell of a rotting, fly enshrouded animal carcass; I've smelt that stench on dead people, and I swear to God in all mighty heaven, I can smell that scent more real in my dreams then I have in reality. Every night at about this time, I find myself sitting at the table and staring out the window. My fading headache and night-sweats have signaled the end of another night time terror. It's time to start drinking again. Some people might think that everybody sleeps at night and work throughout the day-far from the truth. There's lots of life happening down there on the streets, when it's dark and cold. Folks are busy doing all kinds of things. Some are important, others just useless. But Mr. Sandman will watch the world toil tonight, through his only, lonely window. But how do I know he's lonely?—Because I am Mr. Sandman.

Just once . . . damn it; just one lousy night I'd love to have a good night's sleep with no interventions, no interruptions, and no nightmares. No headaches and sweating. No tremors; no tears. But now it's become another night of watching the streets and the people who live in that night-strife. My neighbors think I'm demented. They've heard me waking up screaming, and sometimes they even pound on my door yelling loudly, "Is everything all right?" Or sometimes they'll yell, "What's going on in there?" Other times, they threaten to call the police. Most times, if I ignore them long enough, they'll just go away. The people who live in these apartment rooms usually don't stay very long. Most of them come and go like borderline vagrants—Modern-day nomads.

I keep hoping that the doctors will find a cure for me, but I'm saddened that in all probability, they won't. They tell me there's a hyperactive spot somewhere near the stem of my brain. I've taken many different drugs for my condition, but none has helped. They told me that the hot spot is near the fight-or-flight part of my brain and somehow, the two areas get mixed signals. When I'm awake, I suppose other parts of my brain take control and that makes me feel normal. However, and extremely rare, my hands will sometimes tremble and my irritability doesn't completely go away. I've substituted my medication with drinking and smoking as the only way of surviving my lonely nights.

Jobs are hopeless for me. My caseworker will account for that. My only source of legitimate income is from the government; but it doesn't pay enough. It won't pay for my beer or my regular fifth of scotch; nor does it pay for my cartons of cigarettes. I have to find my own money for those. Probably many people wouldn't believe some of the things I do for extra cash; let me just say-I go for walks in the night.

It's time for another beer, another shot of scotch and a cigarette, and maybe even another three aspirins. My home is a small apartment on the seventh floor in a building where rats live in the walls and old winos die in the hallways and stairwell. My apartment is small. The bathroom is a little cubical which almost smells as bad as a whore's barroom shitter. I wish that it had a door instead of a drape. The moldy, peeling and damp bile-jointed pipes' sticking out and down the walls give me the impression of prison. I try and keep it clean with Lysol, but by the end of its cleaning, it retreats to its same old smell. My bed lies on the floor about four feet from the table and my only window beside it. The kitchen is fixed in a corner with a small two-burner stove, refrigerator and sink. Oh . . . and

I almost forgot about the dry-rot wooden floor counter and the small cupboards above.

There's an uncovered light bulb that hangs from the center of my kitchen with a pull string, and a miniature lamp beside my bed on the nightstand. My clothes are suspended from a thick and bent, wooden dowel attached to the wall behind my only entrance. I keep my bag of tricks in the last bare corner of my home. Most nights, I keep my kitchen light off to hide my reflection from the window glass, and to also see clearer outside. From the streets down below, my head looks like a little ball with no distinguishable features. I know this because I live on the seventh floor, and I've seen people through their windows up that high before-little tiny balls.

I keep telling myself that things could be worse, and I know this to be true-my window has shown me so. It was on a Saturday night a couple of weeks ago that I witnessed a security officer die in the entrance of a small parking lot. A gang of street punks beat the living shit out of him while he surrendered his life for a late model Ford. Stupid bastard . . . but hey, everybody's got a job to do, and a damn good job he did too. I would have called the cops, but without a phone, I was helpless to do nothing but watch. I'll never forget that night.

It was the night of the spider dream, which woke me up that time. I don't have the spider dreams too often but when I do, they torment the hell out of me-I hate those damn things. My nightmares are in and of themselves bad enough, but more cruelty is spent on me while I sit alone having my dreaded flash backs-my minds self-prescribed dream therapies.

I don't have a television and so the only form of entertainment is the window and my radio. Sometimes I listen to the news, while other times I listen to the late night talk shows; and then there are the rare occasions when I listen to music while looking out my window. This is my nightly reality; drinking scotch, chugging beer, smoking cigarettes, watching out the window and listening to the channels on my radio. If I start with a full bottle of scotch, I'll eventually squander my night in a dumb bum's stupor. I have a .38 caliber handgun with a silencer attached in my bedside table. I couldn't count the number of times that I've suckled on the end of that barrel, only to remove it from my mouth because the Bible forbids suicide, or maybe I'm just a coward and afraid to die. However, when the scotch intensifies my outrageous behavior, I'll go for a night walk with my gun and a disguise, which is hidden in my bag with camouflage.

About three nights ago, after drinking scotch and feeling courageous, I disguised myself as an old man with a gray beard, fake teeth, gray tangled hair and ragged clothes. I hid outside against a wall in the shadows of a fancy nightclub's parking lot. I carefully watched the wealthy people walk out, or more accurately, stumble into the parking lot. I waited patiently for just the right person and moment.

At last, my victim appeared by himself. He was intoxicated, mumbling and digging in his pockets for keys. I carefully and quietly approached him from behind. With all my might, I hit him on the back of his head with the butt of my gun. He fell forward and hit the pavement hard-I knocked the poor bastard out. I quickly searched his wallet and took only his cash. I did this shit one time in four different lots. After the evil deeds were through, I'd take off my disguises and hide them in my pockets, which I had sewed inside my coat sleeves. As always and only when I knew I was safe inside my home, I counted the money. That night I made a little over eight hundred bucks. Please don't misunderstand me, but I have to do what I do to survive, and never once, have I ever killed anyone.

Later on in the afternoon, in the safety of daylight, I'll go to the store and buy some more beer, scotch and cartons of cigarettes. I'll also have to buy more food, toiletries and a few other items including a porn magazine or two.

In my early and more naive days, I actually tried to be a character comic, with my bag of magic tricks and stupid faces. I tried to impersonate weird people, but after six months of that nonsense, it became so very apparent that my routines were nothing but rubbish and tasteless bullshit. People always laughed at me and not with me. I flopped on my ass harder than a hammer slamming down on a cold thumbnail in the dead of winter. I must have been out of my mind to try such a moronic stunt. I knew deep inside that I wasn't cut out for this type of foolishness.

I should have known better-after a short time in that ridiculous business, I quickly came to realize that this shit wasn't for me. My life was filled with nightmares and unhappy circumstances. I couldn't wear a disguise and say funny lines because I was trapped in a sad life that bordered on insanity. My life has become unraveled all because I developed this evil wicked sickness of relentless nightmares when I'm sleeping.

Buy the next day I was still awake, and so I decided to spend some money and go shopping. First thing I did was to buy my supply of alcohol and cigarettes. I always purchase these items at *Billy's Party Store*. I've known Billy for about three years. He knows of my condition and is sympathetic, but I told him to tell nobody-not even his wife. We'd talk a little bit, mostly about me and other little things, and then I'd leave Billy's to return my apartment with my booze. As soon as I walked into Billy's store, he'd say, "There's my friend; Mr. Sandman," and then he'd put his arm around my shoulder and give me a hug, "How have you been, Sandman?"

"Nothing's changed, Billy, I'm still having those damn nightmares and all the other shit that goes with it."

"Oh, that's such a shame. I feel sorry for you, Sandman. I wish there was something I could do-when's the last time you've seen a doctor?"

"A few months ago, they stuck a needle in the back of my head and drew some fluid out of my brain. They told me that they're working on some new chemicals that might stop my dreams from driving me goofy."

"Well, that sounds encouraging, wouldn't you think?"

"I don't know, Billy, I've just about given up going to the doctors-they're not helping me."

"But, Sandman, you have to give them time . . ."

"I've given them five of my fucking years and they still know nothing!"

"Sandman, now I want you to calm down and get yourself relaxed. I didn't mean to get you all excited, I'm sorry."

"No Billy, I'm sorry, it's okay, and I'm just not sure how much longer I can take this crap."

"The doctors will find something soon, Sandman, I'm sure of it. They know what they're doing. What you have is probably very rare.

Things like this always take time. Don't hate the doctors Sandman, pray for them. They're on your side."

"You're right, Billy. You're always right. Please forgive my anger."

"Forget it my friend, and I can appreciate your anger. I just keep wishing there was something I could do for you. Would you like to come over later for dinner with my wife and I-she's a hell of a cook. Tonight she's making a pork roast dinner. I think you would enjoy it Sandman. Would you like to have dinner with us later?"

"Thanks for the invite, Billy, but I'd rather kick back tonight with a beer or two, and then try and catch some sleep before my dreams turn in on me."

"You know, Sandman, you should slow down on the alcohol-it's not good for you."

"I'll be okay, Billy . . . maybe sometime in the future. Maybe if these dreams go away, but not now, not today, not tonight. I have to drink."

"I worry about you, Sandman. I would feel so bad if something should happen to you. Just remember, you know where I work and live. And if you need anything, you come see me. I'll pray for you as always."

"Thanks Billy, I'll catch you later."

Billy always lets me use one of his little shopping carts to transport my beer, scotch and cigarettes back to my apartment-three blocks away. I used the front elevator to get my nightly boxes of sin, to the seventh floor. I bought three cases of beer, four bottles of scotch and five cartons of cigarettes.

The total cost, including my 10% discount was $372.54, without the sin taxes. But the government sends me a check every month for $300. Now you see why I have to do other shameful things to supplement my addictions. I freely admit that I'm a lousy bum and thief, but I will not apologize, because I have to do these things in order to survive. That, my friend is the name of the game: staying alive and living through it.

After I put a case of beer in the refrigerator, I stacked the rest in my bathroom along with my bottles of scotch. Most of the cigarettes went in the freezer, while the rest I sat in a kitchen cupboard. Then I returned Billy's cart and proceeded to go to the grocery store to buy some food and other assorted items. I keep a diary of my nightmares in notebooks, so I usually pick up a couple of new notebooks and pens for future use. And I always keep a good supply of over-the-counter drugs: sleeping pills, for all their worth, aspirins, high potency multivitamins, muscle-aching ointment, eye drops, cough syrup, iodine, hydrogen peroxide, bandages, surgical tape and even a small surgical kit. If I should get cut real bad and need stitches, I'll know where to go.

The grocery store is just around the corner and one block over. I was carrying three bags of food and medical supplies, which I had bought from the grocery and pharmaceutical stores, when suddenly the grocery bag tore open and spilled on the sidewalk. The walks were filled with people coming and going but none would help, they were like zombies caught in their own blind spell of their busy nether world. To put it simply, nobody really gives a shit about anyone anymore.

"Here . . . let me help you with that," she said.

I set the other two bags down on the walk and started to help her gather my things and said, "Thank you very much, miss."

"I'm afraid we can't put anything back because your bag is ripped and . . ."

"Could you just watch my things for just a moment?" I pleaded. "The grocery store is right around the corner and if I run, I can be back in just a couple of minutes with new bags. Would you watch my things for me?"

"Sure. Why not? I'll just stand here and wait," she replied.

I returned in three minutes. We started to put the spilled items into a new paper bag and when we were through I said again, "Thank you very much, miss . . . you know, most people wouldn't even stop to help."

"Oh good grief, isn't that the truth. My goodness, I saw at least seven or eight people walk right past you."

When we were through talking, I sat the bag next to the others and discovered ourselves standing up and looking at each other.

"My name's Darlene . . . *Darlene Osmond*," she said holding out her hand.

Not knowing what to say, I quickly looked around and saw a van drive by advertising, 'Christopher's Dance Studio'. "Oh . . . I'm sorry. I thought I saw someone I knew," I said pretending and shook her hand, "My name is Christopher . . . *Christopher Lance*, nice to meet you, Darlene Osmon."

As we shook hands I noticed a wedding ring on her finger. The last thing I need in my life is some married woman who cheats on her husband. She was well dressed and extremely attractive. I saw her pretty eyes sparkling from the sunshine. Again, I looked at her wedding ring and said, "Thanks for helping me Mrs. Osmon, but I'm sorry I've probably taken up most of your lunch hour or something, I'll leave you now; thanks again."

"You're welcome, Mr. Lance. You must live nearby," she said abruptly, making me stop and turn around.

"Why do you say that?" I inquired.

"People just don't go walking around juggling three bags of groceries in their arms unless they live close to home."

"Interesting observation," I said.

"Tell you what, Mr. Lance; I work in the Atwood Building just down the block. My office is on the twenty-third floor of Martin & Kelly Publishing. Stop by sometime, say about 11:30 and we'll go for lunch-my treat. Gotta go—bye-bye."

She walked away looking back once and winked her eye. I waved good-bye, picked up my grocery bags and continued down the street. For the first time in a very long time, I actually caught myself smiling-but for what, I really didn't know. At the end of the street was my dilapidated apartment building. When I was in my room, I sat the groceries down on the table and stared out the window. Standing in one spot, I thought to myself that was the first pleasant thing that's happened to me since I moved to this big and busy city.

After a few minutes, reality set back in and I began to put away the groceries. I popped open a beer and grabbed a fifth from the bathroom, and then sat them down on the table. As I finished putting the groceries away, I stopped and said out loud, "Pork roast dinner is now being served," and then sat down at the table next to the window and began drinking my scotch and beer. I lit myself a cigarette and poured another glass of scotch. I took turns with my drinks. First, I'd take a gulp of scotch, and then a swig of beer and chase it all down with a few drags on my cigarette. This was my drinking routine, and tonight, this was *my* 'pork roast' dinner. I turned the radio on and tuned it to the local news channel.

I heard the announcer talking about a man who was robbed at a plush nightclub and received a blow to the head requiring fourteen stitches. I said a little prayer, and then continued to listen to the radio while turning my head to look back out of the window.

At about nine-thirty, I decided to try and eat something, and then go to sleep. My body felt haggard and my mind was drunk. I turned the radio off, went to the refrigerator, and peeled three slices of bologna from its package and stuffed them all in my mouth. After my small, cold rolled meal, I stripped off my clothes and lay on my bed. As I always do about this time at night, I wondered how many hours of sleep I'd get before Mr. Sandman wakes up and spoils my sleep with his nightmares. I closed my eyes and the last thing I remembered was Darlene's gorgeous face.

"Aha! Oh my God! Don't chop off my head! Please, don't look at me like that! Don't do that with . . ."

As I rolled out of bed and onto the floor, reality started to enter my mind again. I was sweating badly and my head was pounding hard. I stood up and grabbed my bottle of aspirins from the kitchen counter. My hands were shaking badly but I managed to open the bottle and spill out a pile of pills. I scraped five off the counter and flushed them down with a warm beer that was left over from earlier. I retreated to the table and lit a cigarette.

"The decapitation dream . . . damn . . . goddamn it, why me?"

I laid my face in the palms of my hands and started crying. After the tears, I got a cold beer and looked at my watch, two-thirty in the middle of night. I turned the radio on and switched the channel to a talk show and then began my vigil at the window. My mind started to relive the nightmare and I began to wander into disconnected segments while my eyes stared out the window focused at nothing. After I pounded down two bottles of beer and a tall glass of scotch, I decided to go for a walk. I shuffled over to the corner and opened my bag of tricks and disguises. This time I used the bald-headed and thick eyeglass look. My watch read a little after three. I dug around my pile of dirty clothes and dressed. I put on an old trench coat and slipped my gun into the pocket. On my way down the stairwell I stumbled over a bum sleeping on the steps.

"You son of a bitch!" I yelled. He opened his eyes fairly wide and stared at me wondering if he should stay or go.

"Don't just lay there ass-hole, get the hell out of here!" I yelled again.

"I ain't got no place to go mister."

"Ah, fuck it, fuck you," I said viciously and close to his face, and then proceeded down the stairs.

I left the building and started to walk down the street, and although I was probably drunk, my confidence told me to go on.

I needed more money; plain and simple. As I proceeded down the sidewalk, I was careful to watch my surroundings and be on the alert for any sudden surprises. If I walked by an alley or a darken street, my peripheral vision, like a new sixth sense, would immediately kick in. And I'm always on the defensive because at this time of the night, anybody whose out here knows that everybody's prey. It wasn't long before I saw someone at a distance walking toward me. Whenever I meet someone who might be looking for trouble, I always play stupid. This little act always tickles me. No matter who my challenger is, or how smart he thought he was, it always had the same effect: it takes their defenses down. I sat on a bench and watched the person grow closer. I nonchalantly looked around me to see if any other people were watching. The person, who turned out to be a man, came closer still. I lit a cigarette and then watched and waited. And as I thought he would, he stopped in front of me.

"You got a cigarette?" he asked, while he looked around vigilantly.

"Sure," I said acting a little half-witted but consciously realizing where I wanted this situation to go. I pulled a pack out of my shirt pocket and shook it, pretending to be nervous, until a cigarette popped up. He snatched the cigarette without even saying thanks.

"You got a light?" he said in a very intimidating voice, which I chose to ignore.

I stood up to light his cigarette, and sized him up. His hair was combed nicely and he dressed reasonably well. But now it was time to ask myself that silent question again: "Is this guy carrying a gun?" My answered was instinctively, yes.

"Later, chump," he said and began walking away.

"No problem my friend," I said in a manner to invoke him.

He stopped, then slowly turned around and said, "Do I know you?"

"No," I simply responded while looking dead straight in his eyes. Reverse intimidation always works, but I have to add, it can be a deadly game to play.

"Then what did you call me?" he asked.

"Nothing. I just said, no problem my friend," I smirked.

"Do I look like your fucking friend?" he responded harshly.

"Look, mister, I'm not looking for any trouble. You asked me for . . ."

"I don't give a shit what I asked you for," he said as he started walking back toward me, and then stopped about three feet away. I slowly put my hands back in my pockets.

"I don't like your condescending attitude; I don't like you and that bald ugly head—now give me your money," he demanded as he stepped too close.

"But I don't have any money, sir . . . I don't even have a . . ."

"Bullshit!" he whispered close, and then spat on my face.

"Give me your money or I'll gouge those fat fucking eyes right out of your head," he said as he unmasked a large knife.

Hum . . . I thought to myself, I was wrong. This idiot doesn't carry a gun. I breathed a little sigh of relief.

"Do you do this for a living?" I asked him while slowly taking my glasses off, and then started laughing.

"Don't play with me motherfucker. I'll give you one last chance; now give me your goddamn money."

"Let me see if I understand this," I said, "if I don't give you my money, you're going to potentially kill me with that knife?"

"If that's what it takes," he said while making a couple jabbing motions.

"And what about you, my friend, what if I should decide to kill you instead?" I replied smiling.

"Yeah, right. What are you smiling about old man! You think you can take me out? Come on . . . get real, you old son of . . ."

Before he could say another word, I surprised him with a high power kick to the throat. He fell backward losing his knife and laying on the pavement gasping for air. I stood over him with my gun pointed to his head. "Listen to the prophets on the streets-don't ever bring a knife to a gun fight. And now you, my friend, you may give me all of your money." He reached into a pocket and handed me a wad of bills. "You know what?" I said to him, "I prey on fuckers like you. And you know what I always give them when I meet them, pecker face?" I said softly laughing.

"I don't know what? No wait! Come on, mister, please don't hurt me; I've got a wife and four kids, man!"

"Oh! I'm sure you do. And I'd bet you have just as many wives as kids, maybe even a girlfriend or two, but you haven't answered my question."

He stiffened his back and went into a momentary spasm, shaking and pissing his pants, and then he suddenly asked, "What, sir. What do you give them?"

"A trophy. But now I don't want you to go prideful on yourself because you're not the first. It's called the gimp award. Every time you walk on your gimpy leg, you'll remember it was I, not you, who won this little battle. Further more, and here's the best part, you're retired now." I moved my gun toward his knees. "Eenie menie miney moe, which knee cap should I blow."

"Please, sir, don't shoot, please . . ."

I took aim at his right knee, which he seemed to favor, and with a small squeeze on the trigger; there was a little pop that made a big mess. His knee shattered into a hundred little pieces. Bone and blood splattered all over his pants. His lower leg was still attached but with only a few strands of muscle and ligaments.

He started crying and yelling, "My leg! My God, you shot my leg!"

"Hey, I'll bet that hurts; dumb-ass. Well, gotta go, hope you have a nice life, *my friend*."

I walked away in a rather expeditious manner, and then ducked through an alleyway. Within minutes, I was far from the scene and headed toward home. And only when I was home and safe, did I count the money. That creep had $1,200 on him, and now it was mine. I put the wad of money in my bedside table along with my gun.

Usually, I go to bed early in the evening even if I'm drunk. I don't know why, but it seems like the best time to sleep-maybe because it's developed into some type of habit, but that's just a guess. It was six in the morning, only thirteen more hours to go before I get another bout of sleep. I poured myself a new glass of scotch and retrieved another cold beer, and then I sat at the window and surveyed the city below. I saw a man running down Steward Street, and a second later, I spotted someone else chasing him. The chasing man came to a stop when a girl, who also seemed to be in the chase, ran to him. I could see them talking and waving with their hands when suddenly, their talking changed to arguing and pushing. Finally, she slapped him.

This apparently really set him off and he began slugging and smacking her around like she were a man. I grabbed my binoculars and studied the pair. To my surprise, the girl was Darlene, and I could only assume that the man was her husband. He grabbed her by the arm and within a few minutes, they were out of my view as he dragged her away. I thought about going down and intervening, but common sense told me otherwise. It was so damn early in the morning-why was she out there and why were they fighting?

I hung the binoculars back on the nail in the wall and sat down at the table. I took a drink from my glass of scotch, a gulp of beer and lit another cigarette. I was getting my buzz back and enjoying every minute of it. It seemed to be my only salvation, which would ultimately lead to my destruction. Drunks usually don't live very long. I drank and smoked, and felt my beard grow.

I spent the entire day switching places between the shitter and the table in a drunken stupor until seven at night. My eyelids were droopy and my brain was soaked in alcohol. I knew it was time for bed. I laid down fully clothed and fell into a deep sleep-at least until Mr. Sandman started to dream again.

"Let me out! Please, oh God, oh God! Damn you all! I don't want to die . . ."

After about six hours of sleep, I was fully awake. As usual, my head was throbbing and my body was soaked in sweat. My heart was thumping and my hands were shaking again. This time it was the buried-alive dream.

After taking some aspirins I snatched my notebook and scribbled down my dream. Although my nightmares are recurring, there's always some variance in them and so I write that in my notebook. I had my suspicions, but now I'm quite sure that it doesn't matter what time I go to bed; Mr. Sandman will always wake me with his nightmares frolicking around in my tormented head. It was about three in the morning again, but today I was determined to get out of this apartment.

After my headache went away, I made myself some coffee and waited for dawn. After watching the sunrise I decided to clean up. I mean, really clean up: shaving, brushing my teeth, shitting and taking a good hot shower. Afterward, I put on my last set of clean clothes, grabbed all my money and headed out the door. I wanted to feel human again.

My first desire was a haircut, which I purchased from a pretty young woman in a sporty salon and was so impressed by the improvement, that I gave her a generous twenty-dollar tip.

My next wish was for a few new clothes. There was a fancy clothing store not far away from the hair salon. I bought three nice pairs of pants, new shirts, a few sports coats, a black leather trench coat and a pair of neat black boots with zippers on the sides. The total cost was $942.58. I left the store wearing a new set of clothes. My last stop was at a nice restaurant for breakfast. After finishing my plate, I flipped a ten spot on the table for a tip and then grabbed my shopping bag and proceeded toward the Atwood Building.

I entered the lobby, walked to the elevators and pushed the up button. It didn't take me very long to find Martin & Kelly Publishing because they leased the entire 23rd floor. Like a lost stranger I went inside and stood at the reception desk.

"May I help you, sir?" an elderly woman asked.

"Yes, I'm looking for Darlene Osmon."

"I'm sorry sir, but she's not in today. Is there anyone else I can call?"

"No, that's okay; I'll stop by another day."

I left the building and started for home. When I arrived at my room, I undressed and put my new clothes back in their plastic covers so they wouldn't smell of cigarette smoke. Then I put on my usual garb, stuffed a dirty pile of clothes into a duffel bag, including my nasty bed sheets, and left to do my laundry. After returning from the Laundromat I hung my clothes and made myself a clean, bleached and Lysol disinfectant bed, and then sat at the table with a nice rewarding cold bottle of beer.

However, it wasn't long before I had drunk a half bottle of scotch while listening to the radio and staring out the window. It was going on eight-thirty and the night was just getting started. I should have been going to bed, but I wasn't the least bit tired.

While listening to the radio, I heard a local news story about a man getting shot and robbed last night. The newsperson gave a description of an old, bald-headed man with thick eyeglasses who shot and robbed him. I finished a gulp of scotch and smiled. I knew from the very start of our confrontation that the victim was going to be him.

Across the street and close to the intersection of Steward and Browne was a small factory. There was a man driving a forklift truck from one ramp to another and yet another. He had stacks of boxes on a pallet and was transporting them from one ramp to the others. Suddenly out of nowhere, I saw a car come into view screeching around the intersection and then slamming into the forklift. After the collision they bounced away from each other like billiard balls.

The driver of the forklift was pinned under his machine. I saw three people run from the car. The driver jumped a fence and hid in a dumpster while the other two ran off down Browne Street. I stood up just in time to see a police car zoom through the intersection with its lights and siren blaring and going pass the scene of the incident; apparently unaware that the car had turned the corner. I ran out the door and down the stairs.

My first impulse was to get the guy hiding in the dumpster. I jumped the fence and ran to his filthy hiding place.

"Come on out, mister!" I shouted. There was no answer but I could hear his heavy breathing.

"I know you're in there. Come on out or I'm coming in!"

Still nothing; so I jumped into the dumpster and found myself in a struggle. He was tough, but I was vicious. I finally managed to throw him out of the dumpster with myself on his back. His arm was broken and twisted completely around. I dragged him by his broken arm to the fence and shoved him over. Like a rag doll, he flopped across and landed on his ass in a sitting position. I quickly hopped the fence and made him get up and walk across the street to the scene of the crash. I saw the man under the forklift with a partially twisted torso and a leg on top of his chest.

"Somebody call 911!" I screamed.

"We already have!" someone shouted.

There was a crowd of people around this unfortunate mishap doing nothing but watching this poor man bleed to death.

"Will some of you men hold on to this bastard for the police? He was the driver!"

And with that said, four men hurried over toward me and grabbed the driver.

I immediately ran over to the guy pinned under the forklift. He was indeed, bleeding badly. Blood was dripping from his ears, nose and mouth, and from the waist down. I knew his body was crushed by the weight of that forklift. I took hold of his hand and he gripped mine tightly.

"Hang in there my friend, the ambulance will be here soon," I told him.

"Am I going to die, mister?" he asked me with one eye pointed to the ground while his other to the starry night sky.

"No, no. It's just an accident-you'll be okay. What's your name?"

"Jimmy."

His left eye was protruding, bleeding and as I said, pointing to the side of his head. He had sustained injuries so severe that I knew he was going to die. His blood loss was a definite sign.

"Just hold on to my hand, Jimmy, and you'll be okay."

"I can't see you . . . please don't let me die!"

"Where's the fucking ambulance!" I yelled while looking around the helpless crowd.

Jimmy started talking again, but his voice was weaker and his grip was becoming less. He was going into shock and trembling.

"Are you still there, mister?" he asked with a blood-spitting cough. I brushed back his hair and whispered in his ear, "Jimmy, I'm still here, and I promise not to leave you," and then I gave his hand another squeeze.

"Please, mister, tell my wife and kid I love them. Please . . . promise me? I think I'm going to die," he said with sweat and blood on his brow.

"I promise, Jimmy. You just hang in there and we'll have you fixed up in no time."

"Mister, I can't feel anything . . . I think . . . I think I see . . ."

At that exact moment, I heard him exhale his last breath while his hand let go of my grip. I grabbed him around the shoulders and hugged him as hard as I could.

I cried and said a silent prayer. After a minute of grief, I got up with bloodstains all over me and saw a woman standing next to me watching.

"Lady?" I asked.

"Yes?" she replied wide eyed and holding her hand over her mouth.

"Please tell this man's wife that the last thing he said before he died was that he loved her and his kid. Will you promise me you'll tell 'em?"

"Yes, of course, I promise," she said.

I thanked her and than ran off in the opposite direction of my apartment building and circled around a few blocks before sneaking back into my room. I couldn't control myself from crying. I ripped off my bloody clothes and threw them in the trash and then took a shower. Afterward, I put on some clean clothes and stood by the window looking out.

The ambulance had come and gone leaving only the police with a few bystanders and spectators still on the scene. I sat down at the table, poured myself some scotch, lit a cigarette and watched. The incident had sobered me up. My mind had to disseminate all the events that just happened.

The reality of this frightful display was about as frightening as some of my nightmares. If it weren't for my bruises and scratches, I might have thought it was a dream. I scrutinized the spectacle as it was being cleaned, and after awhile, the people slowly strolled away. The police left with their suspect and the forklift was up-righted and hauled inside while the car was being towed away. The only thing that remained was the pool of blood from Jimmy-the unfortunate victim of a senseless crime.

Eventually, a fire-truck arrived and hosed down the red stains that had recently been the life giving force of a human being.

CHAPTER TWO

I pulled the string that hung from the kitchen light bulb leaving only my bedside lamp on. When the kitchen light went off, the room became a warmer place. The paint chips and cracks were less visible and I could no longer see my reflection in the window. Slowly, I became aware of a noise in the background that was bothering me. I discovered that the radio was still on but the volume was low-I must've turned it down. I turned it back up only to hear the news again. It was all about the shit happening in the city and I was sick of it. I switched the channel to some music-soft music. I chugged down what was left of a beer and went to the refrigerator for a new cold one.

I sat back down at the table, lit a cigarette and gulped down a mouthful of scotch and chased it with a cold swig of beer, and then another puff from my cigarette. After about thirty minutes, I started to feel a little relaxed and a bit more peaceful. As I was looking out the window, my eyes were drawn, as if by some compelling force, to that spot on the street.

After the water had dried, it was as if nothing tragic had ever happened. Jimmy just became another statistic. And unlike famous or popular people, he would be mourned by only a few. I had to wonder if his uncelebrated life had any purpose other than procreation. And then I had to wonder about myself.

After a little contemplation, I skipped my empty glass of scotch and went directly for the bottle. I tipped the bottle of scotch up and drank until it was empty. I got up to go to the bathroom and after I pissed, I seized another fifth and then another beer. I refilled my glass, popped the beer open and lit a cigarette. Staring unknowingly at the silly patterns on the plastic tablecloth, I caught myself thinking again about the most elusive question of life-its purpose. I often think about that question. I have to. And everybody seems to have a different point of view. Many people haven't any answers, but there are many others that say they do. Most people say they find the answer in God. I happen to be one of them . . . well, sort of-except my God is different-but that's not really important. And although I continue to pursue this nagging thought, I think I've found an answer.

Why I keep thinking about it I can't figure out because my conclusion is always the same, even after the many years of mental deliberations. I began by imagining planet earth with no people, but instead, just trees, flowers, majestic mountains, grassy plains, and the oceans and *maybe* even insects-a non-inhabited and beautiful world. Everything is divine and in harmony. No humans or creatures intelligent enough to have any sense of self-awareness; who cannot think, create or destroy. And then I expanded this idea to encompass the entire universe. Suddenly it hit me. A universe with no intelligent life cannot be complete.

For example, consider the universe as the ultimate and most beautiful work of art, (which it is), but with no eyes to see, no minds to think, and no thoughts to conceive; everything then becomes without purpose.

If there were no one to behold, then there would be no reason to compose. And therein lies the purpose of life-to be a witness and to be a part of, and to acknowledge the universe that something, somehow has created. A universe with no intelligent life wouldn't make sense. It would be like a small child who just colored her very first pretty picture. But with absolutely no one around to see, relish and cherish her work, what would have inspired her in the first place? Nothing. She would have created nothing.

Ah . . . to hell with it! I had to let go of these thoughts again because they really troubled me. Besides, who am I but just another person with another improvable opinion about the purpose of life and all the shit that goes with it? Subconsciously, and unaware of my physical actions, I discovered that I was on my fourth beer and one-third of my new bottle of scotch. I turned my attention to the window again. The soft music that had been playing on the radio had turned into a talk show. At first, I was merely listening, but after a few minutes my interest grew, especially about the subject. Then the show began taking calls from the masses.

"You're listening to *The Gray Area*, station WNET 107.3 on your FM dial. Hello, I'm your host, Jeff Kalloway, and tonight we've been talking about genetic science and cloning. Let's go to the phones now. Okay . . . we have a caller on the line, go ahead ma'am, you're on the air."

"Yes, I wanted to ask professor Donaldson when she thinks life begins?"

"Dr. Donaldson, would you care to respond?" the host replied.

Ms. Donaldson responded, "Well, for the majority of people I would have to say; life begins when conception is established. But this is just the definition of meiosis. The question that should really be asked is: does the female egg constitute the beginning of life? After all, it is a living organism. But I would hesitate to call it human life.

Does that seem to make sense?

"Makes sense to me," I mumbled over the caller's voice.

"Okay, then . . ." the host's guest continued, "let's just concentrate on the first cell division; does this constitute human life? I think not-they're just simply two cells. And if there is some unnatural polarity in the egg, than in most probability, the fetus will abort. It's no divine secret; we've studied all kinds of cells. And there are many single-cell life forms too, but do we place a special significance on their lives? You wouldn't call killing an amoeba murder-would you? No. Why should our totipotent cells be any different? They don't know how to *think* any more than an amoeba does?"

"Interesting comment, Dr. Donaldson," the announcer said, and then added, "we have another caller; you're on the air, sir."

"Ah . . . yeah? Hello?"

"Go ahead, sir, you're on the air."

"Yes, all these new ideas about when life begins are making everything too complicated. Let's get back to the good old days when we knew damn well when life began!" the caller announced before the phone went slam-dead.

"Fuck you and your good old days . . . asshole," I told the caller.

"We have another caller, go ahead, you're on the air."

"Yes, I'd like to get back on the subject of cloning. I wanted to ask Dr. Pinkerton about the disaster reported four years ago, at the Karing Institute; they cloned a monster there. Now the debate is whether they should kill it? Doesn't this give you some inclination that cloning may be a bad idea?"

"That was some time ago, relatively speaking, and that was an illegal and unknown experiment which was shunned by the scientific community.

However, I'll have to add this: we retrieved some very useful information from that mistake. But let's not fool ourselves; we've made a lot of progress in this field. And it's becoming clearer to me that cloning will emerge as a very viable part of medicine-its exploitation will be for our benefit. Look at all the diseases we've found already. But I'll agree with the caller, cloning another human being just for the sake of cloning is unnecessary and uncalled for-it would serve no purpose."

"Jesus! Now why'd he have to go and say that for," I replied mournfully.

"Dr. Donaldson," the host said to his guest, "I was fascinated by your discussion of cells, would you please elaborate a little more for us?"

"Of course. Where I was trying to go with that was basically the state of awareness. We human beings are self-aware, and some even believe that animals have this cognitive state of self. There are people who even talk to their plants.

But cells are very small things . . . and they certainly don't *think*-they respond chemically. It's the remarkable arrangement of very large molecules and atoms that make them behave as they do. And if you think about it, our identity, maybe even our soul, is based on a single molecule-DNA. However complicated it is, it's still just another molecule. Now if someone should destroy some atoms in this molecule, are they liable for murder? It's ridiculous. And the same thing goes for a human cell . . . or any other cell for that matter.

"So what are you saying?" the other guest asked.

"Look . . . what I'm trying to communicate across to people is that we ought to consider that, just maybe . . . just maybe life begins at the state of consciousness. When the moment a fetus's brain ignites. But not on a cellular level, where there may be only the division of a few cells. That does not constitute a human being in the truest sense of the word. It has to have a brain."

"You tell 'em lady!" I yelled. "Shit, I think she's right. What are a few cells?

You gotta have a fucking brain to be alive!"

"Alright. We have time to take one more caller before we take a break; go ahead, sir, you're on the air."

"Yes, thank you for taking my call. I think the answer that Dr. Pinkerton suggested hit more to the point. Just the very act of destroying even the first cell of a human being, has prevented its growth and potential to become a human being. Kind of like, quantum mechanics."

"And on that quantum jump, we'll take a break. You're listening to *The Gray Area,* with professors, Mary Donaldson and Samuel J. Pinkerton on station WNET 107.3 FM. We'll be right back after these messages."

"This bullshit is too much for me, I can't listen to this anymore," I commented out loud. I turned the channel to some music, and then began to look out the window again. It was in the middle of February and there were sprinkles of snowflakes falling softly from the sky. Just looking out the window made me feel cold. I saw this particular bum walking up and down Steward Street, holding a bottle wrapped in a twisted paper bag. I'd seen him many times before and I was always curious about what this guy's story was. How, why, and what did he do to become what he is today? I wanted to know what his life story was all about.

So I made a bold decision and decided to invite him up for a drink and some conversation. I put on my coat, slipped my gun in the pocket and proceeded to go meet him. He was sitting against a fence when I found him.

"How you doing there, fellow?" I said.

"I don't have any money, sir, honest. Please, don't hurt me!"

"Relax, I'm not going to hurt you."

"I was just taking a break. Its cold out here and my legs hurt from walking so much, but I have to stay warm. I've got nowhere to go, but I'll leave," he replied.

"Tell you what; how would you like to come to my place. You can get warm and have some beer and scotch with me? Due to a certain little incident, I'm afraid I'll be up all night. I would enjoy some company and conversation-would you consider it?"

"Are you a killer?" he asked looking worried.

"No, no. I'm just interested in talking to you. Come on. I'll help you up and we'll go to my place." I helped him up and we started walking down the street. He apparently was not seriously drunk because he didn't stumble or drag his feet.

"Now, I want you to do something for me."

"Oh, no. You're gonna fuck me!" he whimpered.

"Jesus no for heaven's sake. I just want you to never tell another soul where I live."

"That's it? No weird stuff?"

"Nothing weird . . . promise," I said reassuring him.

"Then I promise not to tell a soul where you live. I won't."

We walked in my apartment building and to the elevator. When we were on the seventh floor, I opened the door to my room and he followed me in. I closed the door behind us.

"Here we are, my lovely home," I said sarcastically.

"Wow, you have a really nice place."

Thinking that I might be doing something really stupid, I had to tell him again, "Don't ever tell anybody where I live, if you do I'll kill you."

"Honest, I won't," he proclaimed.

"Do you know where you are?" I asked inquisitively.

"Yeah. But I'll never tell a soul, I promise on my mother's grave."

"Good enough for me; have a seat." I seized an old folding chair from the kitchen and sat it by mine. I fetched two cold beers from the refrigerator and placed them on the table. I then rinsed two dirty glasses from the sink, filled them with scotch and sat down.

"Please, have a seat," I said again.

"Thank you," he whispered cautiously.

"Feeling warmer yet?" I asked.

"Yes, much warmer, thank you. Can . . . can I ask you a question?"

"You go right ahead, my friend."

"Why are you doing this?"

"I told you earlier, I want some conversation-I want to talk to you," I explained again. He took a big gulp of scotch and looked at me strangely.

"What's your name?" I asked.

"*Scott Sellers.*"

"Nice to meet you, Mr. Sellers," I said extending my hand.

He shook my hand and then said, "Nice to meet you too, sir."

"Please, just call me Sandman."

"Okay," he replied hesitantly.

"How old are you, Mr. Sellers, if I may ask?"

"Fifty-eight. Why do you keep calling me Mr. Sellers?"

"Well, that's your name; isn't it?"

"Yeah, but I haven't had any one call me mister in years."

"That's a shame. Okay, I'll call you Scott," I said as he nodded yes, and then chugged his beer dry and drank his scotch. I got up and retrieved another beer.

"You smoke?" I asked.

"I smoke butts."

"Here," I said and threw him a pack of cigarettes.

"Gee-wiz, thank you, Sandman."

Scott opened the pack, pulled out a cigarette and put it in his mouth. He was looking around for a light when I flicked my lighter and lit his cigarette.

"Here, you can have that too," I said as he puffed to keep it lit. I refilled my glass with scotch and pushed the bottle toward him. "Help yourself, I've got plenty more of that." Scott refilled his glass and looked at me with a puzzled look, and then a smile.

"Talk to me, Scott."

"Talk to you?" he said with a questioning expression.

"Yes. Tell me your story. Tell me all about yourself. Where were you born? How far did you make it in school? Where did you work? What happened to you? Things like that," I said.

"Okay. I was born on a farm not too far from the city. But back then, the city wasn't as big as it is now. My dad was a pig farmer, and when I was a boy it was my job to feed the pigs and clean the shit from the stalls. He also had some chickens and three cows. My mom did the cooking and the house cleaning. She also took care of the garden. I loved my mother, but I hated my dad. Sometimes he wouldn't let me go to school because he said I hadn't finished my work. That really made me mad, and those were the times I'd mouth-off to him and tell him that I'd grow up stupid. And then I'd get my whooping.

He would drag me in the barn and take a razor strap to my back. He told me working was more important than school because school didn't put food on our table. And when I did go to school, the other boys would make fun of me. They called me, *Shitty Scotty*. It was worst when the girls were around, especially Lucy Richy. I really liked her, but I knew she didn't like me."

Scott took a few drinks of his beer and lit another cigarette. He paused for a moment, and then swallowed down some scotch, and all the while, I just kept my mouth shut. "When I was twelve, I killed my dad."

"You killed your father?"

"Sure did. My poor mother was frightened out of her mind. My dad had just given me a big beaten' until I was lying face down in pig shit and bleeding. After I was through crying, I got real mad. And when I got up to shovel more pig shit, I had a change of mind. I took that shovel by the handle and swung it high above his head, and when it came down, it hit him right on top of his bald spot. It bounced and made kind of a *boing* sound. He turned around, grabbed his head and fell to his knees. I popped another one on his forehead. That one split his head open like a melon. He died face down in pig shit."

"Man, that's unreal. So you killed you own father. How did you feel? What did the police do?"

"To tell you the truth, Sandman, it felt great; always has, always will. My mom was mostly worried about me and the farm, and who was going to take care of it. I told her that I would do all the chores; that I would take care of her and the farm. After she quit crying, we put dad in the back of the old Chevy truck and drove him to the hospital. Of course, he was dead when we got there-let me think . . . what did the doctor say? Oh yeah . . . DOA. Anyway, the police came to the hospital and talked to my mom and me.

She told the police that dad had fallen off a loft in the barn and landed on his head. The cops asked me if that's the way it happened and I shook my head yes. That was basically the end of it."

"So, did you take care of the farm?"

"For a little while. I quit school when I was in the eighth grade and worked the farm until I was fifteen-that's when my mother died."

Scott quit talking for a few minutes. He lit another cigarette, and then drank some of his beer and scotch. I did the same. Looking at Scott's swollen and watery eyes, it became very apparent to me that his heart was really hurting and saddened by all this talk.

"Damn, Scott, I'm sorry. I shouldn't be asking you to tell me your life story. I think I'm way out of line here. It's really none of my business . . ."

"No, Sandman, you're not out of line. I haven't talked about any of this stuff in a long, long time. Actually, it feels good to talk about it. Kind of gets it out of my gut."

I waited a few minutes before asking Scott another question. "What did you do after your mother died?"

"I sold the farm for fourteen thousand."

"Only fourteen?

"I know . . . but I was young and stupid."

"What did you do with the money, Scott?"

"I used most of it for a down payment on a small house closer to the city, and then I got a job working on the line for a small company called, *Lewis Manufacturing*-they used to make parts for radio sets. But they went out of business after the television was invented, and due to my lack of education, I was unable to find a new job that paid any fair amount of money. Everyone kept telling me that I needed a high school diploma. I was jobless.

And when the money ran out, I couldn't keep up with the mortgage and all the other bills. I eventually lost my home and sold everything in it to keep myself alive. And now, I've been living on the streets every since. You wanna know something else Mr. Sandman, I never used to drink. Not a drop. But as the time went by, I picked up the habit from the other people on the streets like me. Most of the money I earn now is by begging, but people don't give their money so freely when they know you're just going to buy booze with it."

"So where are you living Scott?"

"I all ready told you, Sandman, I live on the streets. My bed is any place that seems safe. I sleep in abandoned buildings, boxes in alleyways, under steps, and in stairwells. Hell . . . just the other day I was sleeping in the stairwell of this very building, when a man came down the stairs and stumbled over me. Fuck; he scared me to death-I thought he was going to kill me; and he looked like the kind of guy who would do it too. He was bald-headed and wore thick glasses, and this was the strange part; his voice didn't seem to fit his face. Shit-I counted my blessings on that one. He didn't hurt me, but he did do a bit of cursing at me before he left. Man . . . I've gotta tell you, I got out of that place in a hurry. I sure as hell didn't wanna be there when he came back again."

"So, in a nutshell, that's basically your life story?" I asked.

"That's pretty much it, Sandman, not a whole lot to tell. I hope I didn't disappoint you. I'll be right back; I gotta piss real bad."

"First curtain on the left," I slurred.

When Scott walked into the bathroom, I heard his drunken voice mumble out load to himself, "Damn . . . what a dump," and then he raised his voice for me to hear, "Sandman, do you realize how bad it smells in here?"

"Yeah," I had to agree.

I glanced at my watch-about three in the morning-still early. Our beer and scotch were empty so I went to the refrigerator for two new ones, and then bellowed to Scott, "Don't take all night in there."

After taking turns in the bathroom, we found ourselves staring out the window, and occasionally looking at each other in silence. I had another fifth on the table. There we sat, slowly drinking, thinking, smoking, talking and sometimes even laughing. And as quickly as the laughter disappeared, the room had gone silent again-neither one of us said a word for more then a minute, maybe two.

Finally, I spoke, "Do you know what kind of shit that goes on out there?"

"Of course I do, Sandman, why do you ask me such a stupid question?

"Watch your words, Scott, everything's been going well up to his point," I said sternly.

"Oh shit, Sandman! I'm sorry; I didn't mean to make that remark that way. I meant . . ."

"Forget it, Scott, we've been drinking-we're drunk. Besides, I'm in no mood to be angry or mad."

"You're a very nice man, Mr. Sandman."

"I'm just kicking back and letting it roll, my friend. And you're a very nice guest, thank you."

Scott was studying the streets below when he suddenly said, "You saw me from up here, didn't you?"

"Sure did, Scott. I've seen you many times walking up and down the streets below. You always fascinated me and tonight, I finally decided to talk to you. For some strange reason . . . or maybe because of something I witnessed-I'm not sure-but I wanted to know why you were down there.

I wanted to understand why and what you've become . . . I hope you can understand."

"Well, now you know, Sandman. What now? I've told you my story. You know I'm a fucking bum with no money and nowhere to go. What now, Sandman? Is there anything else you wanna know about me, or is it just time for me to go?"

Scott was like a lost and frightened little boy swept up into a crowd of strangers. He held his head between his hands and began crying like a baby. He cried loud and hard with his head cupped between his palms. For a moment, I was beginning to think he was going nuts by the way his head was bobbing up and down. Then, I started to feel sorry for him as he wiped away the snot from his nose and onto his dirty pants.

I got up out of my chair and handed him my washrag, and in a drunken stupor I stepped behind him and placed my hands on his shoulders, and massaged. "Let it out, Scott. Go ahead my friend, just let it all out."

After a few minutes, he began to settle down. I tried to help him relax, "Come on Scott, don't lose it, you're with me-remember? Come on . . . we'll just listen to some music and talk about other things-like what's new in the news and other people-stuff like that." I knew in my head that I was dead drunk when I saw Scott had passed out and rolled on the floor. And then everything went black for me as well.

"Ah! Put it out; take it away! It's hot, too hot! Damn-Oh, Damn. Keep it away! No . . . no! Jesus Christ that burns-ah!"

I awoke all balled-up near the table. My shirt was soaked in sweat and my hair was all wet. I opened my eyes and saw Scott sitting up looking like he had just seen a gruesome and violent death. His eyes were bugged out in complete dismay. I sat up gasping and catching my breath. If I hadn't felt so miserable, I would have laughed at Scott's expression; I knew I scared him.

"What happened, Sandman? Are you okay? Why were you yelling? You're all sweaty. Are you hurt or something?"

"I'm okay, Scott-just a little night terror-give me a few minutes and I'll be fine."

"Sandman, you scared the shit out of me."

I peeled my wet shirt off and dried myself with the washrag that was left on the floor. After going to the bathroom, I went immediately to my bottle of aspirins and rinsed down five of those things. My head was pounding and my heart was still beating hard, but I was gradually getting a grip on myself. I half smiled at Scott and extended the bottle of aspirins to him. He also downed a few. When my eyes worked again, I looked at my watch and discovered it was one-thirty in the afternoon. Feeling a little awkward, I felt that I owed Scott an explanation and told him of my disease.

Of course, he was very understanding and sympathetic. But I wasn't sure how much longer I could survive with so little sleep. And it really didn't matter what time I went to bed, Mr. Sandman will always be sure to come and wake me with his gruesome dreams.

Scott and I were both sitting at the table smoking cigarettes and drinking black coffee. About twenty minutes had past since I took my aspirins and now my headache was starting to ease.

"How are you doing, Scott?"

"I'm okay, just a little hung-over, that's all. But that was the best night's sleep I've had in years, Sandman; I wish I could say the same for you. You must've really had a terrible dream."

"I have them every time I fall to sleep."

"That's too bad," Scott said as he stood up. "Well, Sandman, thanks for letting me stay the night, but I guess I should be going . . ."

"Hold up a minute, Scott, what's your hurry? Come over here and stand next to me; I want to measure your height." And that's all I really wanted to do. To my mild surprise, Scott wasn't that much shorter than me. He stood beside my bed looking puzzled. I walked over to the clothes dowel and inspected some things, "Come over here, Scott," I asked again. He slowly walked over and stood beside me. I held up a couple of shirts next to him, and then some pants. I chose a nice matching ensemble. "Here, these are yours, now go clean up on put them on."

"Thank you very much," he said.

"You're welcome. Now . . . there's a clean towel on the bathroom shelf, a razor and shaving cream in the medicine cabinet, toilet paper next to the water closet and a bar of soap in the shower stall; please use them all. I'll take my shower after you, and then I'll make us a good healthy breakfast. Sound good to you?"

"Sounds great," he said.

After I was through cleaning up, it was time for breakfast. Scott was wearing the clothes I gave him and I had to admit, for out dated garb, he looked pretty good. With a clean-shaven face, combed hair and clean clothes, he no longer looked like the bum that he did last night.

"Here Scott, you didn't put your underwear and socks on."

"Oh. I didn't know they were for me." He went back into the bathroom and came out a few minutes later with a big grin, "feels good."

I decided to give him a prior pair of shoes. "Here-try these on Scott."

He slipped one foot in and said, "Hey, it fits."

"Good, they're yours."

"Thank you, Sandman," he said again.

I nodded my head as an expression of; "you're welcome", and then went to the kitchen cupboards to look for food. "How does Spam sound? . . . Hey? I have a can of corn beef hash . . . you like hash?"

"Sounds great Sandman. I know I owe you something? But I don't have any money on me."

"You can pay me later when you're rich." After we ate, we talked.

"Scott, I want you to know that if you ever need anything, you can come and see me. You now know where I live, but I feel like I can trust you as a friend," and then I walked over to my bedside table and opened the drawer. I saw Scott sneak a glimpse of my gun and money. I pulled out $200 in different denominations.

"Take this, Scott, and promise me you'll do at least two things."

"What's that?"

"Get a hair cut, and rent a room for the night. You can do what you want with the rest." Knowing that Scott didn't have a coat, I gave him my old one with the pockets sewed in the sleeves. He tried it on. It was a tad bit long in the arms, but when he rolled it up one fold, it looked okay.

"Thanks for all you've done for me, Sandman. God bless you."

"Don't be a stranger, Scott," I said while closing the door behind him.

After Scott had left, I cleaned the apartment. I washed all the dishes, cleaned the bathroom again and changed the sheets on my bed. Scott's old clothes and shoes were lying by the plastic trash bag. I picked them up with a dirty fork and inspected each one. They were even more unsanitary than they looked-and nauseating.

His pants smelled like he had shit in them a month or two ago and his shirts were layered and badly soiled. The soles on his shoes were worn through with duck tape on the inside and his socks were nothing but a pair of worn mittens. I threw all his shit in the trash and gathered up all my dirty clothes, including the soiled bed sheets, and put them in my duffel-bag for the Laundromat. Before I left, I opened my money drawer and scooped up the rest of the bills and counted them: $98.

My money was getting low and another government check wasn't due for about two more weeks. When I left the apartment, I made a stop at the trash dumpster and threw away the nasty refuse, and then I was on my way to the Laundromat.

When I got back home I made a clean bed and hung my washed clothes on the dowel, and then sat at the table to rest for a bit. It was six-thirty in the evening and I was dog-tired. With a little encouragement from myself, I laid down on my bed. After last night's lunacy and the rest of today, I was drained. I had no more energy left in me. My eyelids felt heavy and the pillow felt nice. As I starting to doze off, my last thoughts were; I may be a bum and a thief, but at least I'm a clean bum and thief.

I woke up screaming like a demented man who just escaped from an institution for the psychopathically useless. Immediately, there was pounding on the door, "Shut up in there or we'll call the police."

As usual, I ignored the voices on the other side and sat quietly for a few minutes until I heard footsteps walking away with mumbling voices fading with each disappearing step.

"Damn, damn, damn!" I cried out loud. These bloody nightmares have got to stop. I looked at my watch and it read one-thirty in the morning.

After wiping the sweat off my brow, I swallowed some aspirins, walked unsteadily to the refrigerator and pulled out a beer. My hands were shaking and my heart was thumping and fluttering making it hard to catch my breath. I went into the bathroom and looked into the medicine cabinet where I found an asthmatic inhaler. After a few inhales, I hung my head and bent down toward the sink thinking I was going to puke. My arms and hands were quivering so much that I had to clutch the sides of the bowl. Now I was light-headed. After a minute or so, I thought it best to sit down because my legs were feeling weak and starting to wobble.

On the way out of the bathroom, my hand grabbed a fifth, which I placed on the table next to my beer. After that unbearable nightmare, I almost felt totally out of control. And I was completely unaware of how long I had been screaming and screeching through the walls. It could have been two minutes or two hours for all I knew. Quietly, in that dim lit room, I was gaining control. Although the kitchen bulb was off, my bedside lamp was on, and it began to give me a warm feeling of security and serenity. I lit a cigarette, popped my beer and looked out the window.

There were no cups or glasses on the table and so I took a gulp of scotch straight from the bottle. Mr. Sandman was awake now, and ready to scrutinize the dark and cold night from out of the window again.

As I had said before, from my window I could see the intersection of two streets: Browne and Steward. There wasn't a whole lot of action going on at the moment, but when the bars let out, the streets would become busier.

I saw the warehouse where Jimmy died. The new forklift driver was doing the same things that Jimmy did. I looked at the small parking lot where the security officer died. There was a new officer standing down there probably unaware of the tragedy that befell his previous employee. I wondered how long he was going to last-such a thankless job. One would really have to be desperate to take a job like that. He had to be married with kids.

These streets are very bad at night because it's the south part of the city where the gangs, junkies and hookers hang out. The south part of the city is were the poor and homeless live. I looked down where the security officer stood and saw him talking to a man of the streets. Quickly, I grabbed my binoculars and studied them closely. I saw the man hand the officer something that did not elude me. It was money. The security officer took the money and walked into his little four by four offices. I saw him take a key and then lead the man to a car parked near the back rows.

He unlocked the door and the strange man crawled in and laid down. Hum . . . this security person isn't so stupid; he's got a little car motel going on the side.

I turned my head in the opposite direction and saw the skinheads leaving the *Hot Rocks Bar* down the street. It was a rather large bar where the unsaintly come to drink while they listen to the live neo-Nazi music-if one considers that music.

After closing time, they all started stumbling out in singles and groups yelling, barking and screaming all kinds of worthless and ignorant shit. The most atrocious of harmonious voices I heard, were the joy-filled words: "Heil Hitler". Most of these assholes were sporting shaved heads with swastika tattoos along with patches on their leather jackets including skulls and cross bones and other vulgar stuff. The women looked just as mean and dumb as the men. I wanted so badly, with the most purest evil in my heart, to go take a walk down there and shoot all of those sons of bitches and worthless bastards right through their throats. All they want to do is spread hatred and self-imposed anger. They hide under the veil of freedom, but the truth be said; they're nothing more then good-old, homegrown terrorist.

I finished my beer and grabbed another cold one and a glass for my scotch. I poured my glass full and took a deep swallow followed by a big chug of my beer. I lit another cigarette and puffed away. In the back of my mind, I knew that my lifestyle was going to kill me if something didn't change.

But then, I thought how lousy my life was anyway and so I figured, screw it. Turning my head back to the window, I saw the Hitler-Ass-Kissers dispersing and going in different directions. The crowd around the bar was starting to thin out until there were only three or four people left. I was beginning to feel good from the scotch and beer-as a matter of fact, it was time to take a walk in the night.

I got up and dragged my bag of gags to the table and started to rummage through it. Ah . . . this may do the trick. I pulled out a mustache and sideburns along with a braided ponytail wig. Looking into the bag, I also decided to use a hippie-looking set of oval green glasses with wired rims. Before putting on the disguise, I finished my glass of scotch and bottle of beer. I wasn't drunk-just courageous. The wig, mustache and sideburns were tinted to give me a slightly redheaded look. Of course, I used professional make-up and glue to attach the sideburns and mustache. I put on a pair of my new pants and shirt along with my shoes and leather coat. With all the disguises on, it gave me the appearance of someone who might have money, and that was the look I was going for. The last and most important item was my gun, which I took out of the drawer and put into my pocket. As always on these little escapades, I snuck down the stairwell.

The best place to find money is where the money is and it certainly wasn't where I lived. The west side of the city is where the wealthy people live. About six miles away down Steward Street is the beginning of the west side of town.

Stores are cleaner and bigger, sidewalks are wider and better, apartment buildings are more secure, cars are newer and the people are richer. There are no bums on the west side of town.

However, and like myself, thieves like to visit there because they're more likely to find someone with money. But there is a funny aspect to this part of town-it happens when a thief tries to rob another thief. Both of them have no money and they end up killing each other for nothing. Usually, the two thieves are amateurs because they don't know how to spot the right victim. I like to call these fools: Bandit-Yahoos. Surprisingly, it happens more often then one would think.

There's a fairly large 24-hour shopping mall that I have visited once before. Most of the time, I'm forced to walk a few blocks before I can catch a cab. And when I choose a victim, it's always a man. I never fuck around with women. I don't like to frighten them. They may be pregnant or have children with them. I do not harm women; besides, most women seem to use credit cards, whereas men usually carry cash. I don't steal from men after they come out of the store, but instead, just before they go in-before they spend their money. I like to crouch down between two cars behind an empty parking space and wait. I always know when the moment is right.

My preferred stakeout is in the middle of the parking lot for obvious reasons. I was crouching between two cars and observing my surroundings, making sure that no one could see me when finally, a car pulled in and parked at the spot I was hoping. I hid one row behind the car. As the car door opened, I saw the shoe and the pant leg of a man step out. I slowly stood up and started walking toward the car at a leisurely pace.

I pulled the gun from my pocket and just when the guy started to stand up and out of the car, I hit him on the back of the head with the butt of my gun. Before he could hit the ground I grabbed him and gently set him back into the seat of the car. Quickly and gracefully I grabbed his wallet. After I took his money out I tossed his wallet back in and tucked his feet in, and then quietly shut the door.

If you give them a good wallop, most people lie unconscious for half an hour or more, so there's plenty of time to do another one, or maybe two more before I go. I decided to try two more hits. I did exactly the same things that I did before. When I shut the door of my last unfortunate prey, I nonchalantly walked away and headed for home.

When I made it two blocks away, I turned into an empty driveway and took off my disguise and hid it inside my new leather pockets. After my taxi ride, I was back home with a wad of money and threw it on top of my bed. After changing and putting everything away, only then did I count my prize: $1,087.

After paying the cabbies, and including the money I started with, my sum total was $1,180. I was financially set for a while. It was five o'clock in the morning and the darkness would last another three hours or more. I snatched a cold beer and sat at the table with my empty glass and bottle of scotch. I lit a cigarette and poured my glass full.

I reached for the radio and turned it to a talk show, and then I took two big mouthfuls of scotch and nearly downed my bottle of beer. I glanced out the window and studied the streets with thoughtful eyes. One of these days when my luck runs out, the cops will catch me or a stranger will-in which case I'd prefer the cops. However, I disposed of that thought real fast and celebrated my successful night. I spent the rest of the early morning hours drinking, listening to the radio and looking out the window.

At about eight in the morning, I saw the security officer waking his clientele and escorting them out of the cars and parking lot. There were at least fourteen bums leaving the lot. They all left in different directions, including to my surprise, Mr. Scott Sellers too.

After all the robbing and drinking I'd done before the sun started to peek over the horizon, I surrendered my body to bed. It wasn't until about one in the afternoon that Mr. Sandman cursed me again. After some time, the nightmares were through and the heavy sweats were dry while I still sat at my table and cried. I had too much to dream last night. This time it was the eye dream-one that I haven't had in a long time.

I felt tired and debilitated from the nightmares, drinking and lack of good sleep. I looked at the calendar and noticed that the first day of spring was only about a week and a half away. At the end of the month I'll have to see my caseworker and doctor. $1,180 says it's going to be the same old shit but with a different smell.

CHAPTER THREE

Today I decided to clean up and feel human again. After I was clean and dressed, I made myself a good meal. I fixed a few baked potatoes, some fried chicken breasts and mixed vegetables. I'm a pretty good cook when I set my mind to it, and the dinner tasted great. It's been quite awhile since I've cooked a dinner meal. I knew my body needed it and besides, I didn't want the food to go to waste. A little while later, I felt a desire to make some changes to this dump. I decided to go to the department store and buy something new. Maybe a new tablecloth, a curtain for the window, or just something new and different for a change.

The sun was shining and the wind was calm, but there was still about an inch of snow on the ground. While I was walking and shopping, my eye caught some jigsaw puzzles stacked nicely on a shelf. I can't ever remember putting one of those things together, even when I was a boy. I bought one that was a picture of a famous city, and another that was a prehistoric picture of and old spiny tree poised in front of an eerie sunset. On my way from the cash register, I saw some decks of poker cards and purchased a pack.

When I returned, I hung the curtain over the window and was kind of a surprise, that for something so simple, it sure made a big difference. It created an illusion of less poverty. The new plastic tablecloth was the color of the sky, and its new plastic smell helped take away the occasional whiff of order that came from my bathroom. I am not a filthy bum and that bathroom really pissed me off. I had to let the thought go because it was starting to make me mad. I focused my energy away, and then began to wash all the dirty dishes. Afterward, I grabbed the puzzle of the city and spilled all the little pieces onto the table. The thought of putting these puzzle-pieces together made me feel a little good-humored because it felt appealing. On top of that, it was something that I could look forward too. I felt that it would give me a feeling of satisfaction if I could complete it.

I was about an hour into it, when someone knocked on my door. I sense a touch of anxiety because no one has ever knocked so politely on my door before. And there was no good reason why somebody should be knocking at my door. But a thought suddenly occurred to me that it was Scott. I listened quietly for a few seconds, and then the knocking came again, but with louder thuds. After a few seconds, again there was the knocking. I slowly got up and walked toward the door.

"Who's there?" I said rather loudly.

"It's Darlene; Darlene Osmon," she said in a heighten tone.

I was flabbergasted. I turned the lock and opened the door and sure enough, it was she. "How did you know where I lived?"

She hesitated for a moment, and then replied, "I followed you."

"You followed me?"

"Yes, I hope you're not mad."

"No, no, I'm not mad at you; but I'm confused. Why would you want to follow me in this neighborhood?"

"I followed you because I need your help, Christopher."

"When did you follow me?"

"About three days ago after an early morning beating. I didn't go to work that day because of my looks, and since then, I've been calling in sick."

"So it *was* you I saw on the street that morning."

"You saw me?" she exclaimed, "Why didn't you help me for God's sake?"

"I don't like to get caught up in domestic matters; they can prove to be quite deadly; even the police hate those situations," I responded with a slight attitude, "but please, go ahead with your story."

"Well . . . later that day, I snuck out of the apartment and went for a walk to gathering my thoughts, and size up my life. In reality, I was taking a blind walk; worrying about my unhealthy and unfortunate marriage that I felt hostage to, and how I might escape. I was hungry, but didn't have enough money for food; so instead, I went to a coffee shop across the street from a restaurant and had a donut and coffee. By sheer accident, as I was looking out the window, I spotted you leaving that restaurant. You were wearing nice clothes and were dressed handsomely, and I remembered how polite you were when you spilled your groceries that day. Except for only a few people at work, you were the first person I've met since we moved here; four weeks ago. I felt comfort in just seeing you again, as if somehow, and I apologize for saying this, you were like a new friend. Now I feel embarrassed."

"Well I should think so," I said jokingly, and then smiled. "So you also know that I went to your business."

"That's what made me believe that you were interested in me."

"Yes, yes I was. I'll admit that I was curious, but tell me something Darlene; why were you and your husband chasing that man and arguing on the streets so early in the morning . . . why?"

"I know it was early, I was going to work. I was outside my apartment waiting for my cab, when a man ran by and snatched my purse. It frightened me at first, and then made me angry. I immediately ran after him. After about six or seven blocks, I caught him and demanded my purse back. He was old and scared and shaking with a hand full of bills.

He had thrown my purse away while he was running, but not until he was able to grab my money. He had most of it in his left hand, and very willing to give it back; which he was doing when my husband saw us."

"So why was your husband beating you?" I said.

"Oh, let me finish. My husband was out on the streets with his gangster buddies and as crazy as it sounds, by coincidence, that's when he saw us. As the man was giving me back my dollars, my ass of a husband somehow got it into his head that I was prostituting, exchanging money for early sex. He then started to chase the man, and when I caught up to him and tried to explain; that's when he started slapping me around. I could tell he was high on something. But I had no idea that we were on your street corner. He grabbed me with a strong-arm grip, and then took me back home where he beat me again. If he knew I was here talking to you, he'd kill me and you too."

"So what's the deal with your husband and you? What are you?—Fucking stupid for marrying such an ass!" I bellowed.

"Yes I am!" she yelled back, "He turned bad just before I married him. I married Vern while I was in college in my junior year. I was very insecure then. My parents died in a terrible car crash after I had just turned twenty.

My grandparents are also dead and I have no brothers or sisters . . . and I hadn't seen my aunt or uncle for over fifteen years-I don't even know what they look like or where they live. I have no family or friends."

"What about when you were in college? You had to make a few friends there . . . didn't you?"

"Only when I was in school, but when I got home, Vern really never let me go anywhere except to work. And he was still suspicious of me and said I was a whore."

"That should've been your first sign that something was wrong."

"In the beginning I didn't think much of it. Besides, we were married and suppose to be committed. When I first met Vern he was always nice to me. He helped me out with my bills, food, and college expenses. He didn't go to college, but instead, had a fair job working as a private investigator and had a partnership. He made good money until I graduated. After I finished college with a bachelor's degree and a major in literature, I started working as an intern for a small publishing company. And that's when he started to really turn mean to me. Sometimes he would hit me for even the smallest things. He told me that he quit his partnership, and was going to do P.I. work on his own. I told him that I wanted to move here when I was offered a job at *Martin & Kelly Publishing*. Vern didn't like the idea, and we argued for weeks over my decision to move. But I told him I was determined to advance my career, and that he could be a P.I. no matter where we lived. I was hoping he would leave then, but he reluctantly followed me."

"He's a woman beater."

"Oh, God yes! But I don't know what to do?"

"Why don't you just leave?"

"God, I'd love to, but how? He knows where I work and he's threatened me. I'm afraid to call the police because if I do, he said he'd kill me. I don't have enough money to just pick up and go. How do other women do it Christopher? I don't know what to do; that's why I followed you.

"Would you care to sit at the table Darlene? I'd like to offer you a soda or something, but I don't have any . . . would you like a beer, or perhaps a glass of water?"

"I'll share a beer if you'll have one with me," she said shyly.

After I unfolded my spare chair at the table, I motioned politely for her to sit down. And as I was getting a glass from the cupboard and a beer from the refrigerator, I noticed her glances as she secretly looked around.

"So what do you think?" I said as I sat down.

"What do you mean?" she replied as I served her a glass of beer.

"My apartment; how do you like it?"

"Well . . . it's quite manly-a little small, but I'm sure it's cozy, especially at night."

"Please, Darlene, please don't placate me. I'm embarrassed to have you see me like this. I apologized for the way that I live.

"Don't you ever apologize, Christopher, to anyone about your life. It's about the person, not the place."

She was wearing sunglasses and a scarf around her neck, but I could still see some of the bruises through her disguise. "Darlene?" I asked, "Would you take off your glasses so I can see your face."

"It's not as bad as it looks," she said as she took off her glasses. Her right eye was black and blue and swollen, but she told me she could see. Her right cheek was red with small scratches, which I caught at first glance, slanting at an upward direction.

"You've got yourself an ornery little husband, my dear."

"He's not little," she said.

"Is he bigger than me?"

"No, but close."

"I know your skewing the truth, Darlene. As a matter of fact, he's no taller than you . . . probably even shorter. He's bulky and obviously brutish, and from the looks of your face, he has a chubby swing. He's left handed, of course," I said.

"How did you know that?"

"Your right side got the worst. Your husband is left-handed; he favors his left side. He also wears a ring. Those are some nasty little scratches on your cheek-how tall are you Darlene?"

Looking at me intrigued, she said, "five-feet and six-inches."

"Then your husband must be about five-feet-two. Why did you imply he was bigger?"

"I don't know Christopher. I'm sorry. I'm still fooling myself and trying to make him out what I want him to be. He used to be much thinner, but as you obviously saw and perceived, you've described him perfectly."

"Okay Darlene, let's get back to you; what do you want from me?"

"I thought that you might be able to put me up for a couple of nights," she said while looking around my room again and smiling.

"But I can see that you don't have room for me and now I feel embarrassed. I probably should just go, I'm sorry."

"No. I mean . . . stay awhile and talk to me; where have you been living?"

"In a car lot across the street and close to you," she said shamefully and then asked immediately, "What time is it, Christopher?"

I looked at my watch, "Quarter to five, why?"

"I suppose I should just go back home and take my medicine. Vern's probably there waiting for me and ready to kick my ass again. Jesus, this kind of living is lunacy and I have to get back to work; I pray to God they haven't decided to fire me."

"Do you want to live peacefully, or do you enjoy these beatings?" I asked her seriously and showing no facial expressions.

"Why are you being so mean?" she exclaimed, "I'm worried sick." She flopped her head down on the table and starting crying, "Please dear God, help me . . ."

After witnessing that, I knew I wasn't really helping-especially after that particular unnecessary question while showing no emotion. "Okay. Okay, I'm sorry. Truly I am, Darlene. Just give me a minute to think?" I asked with a more soft and sympathetic voice.

"I just need to get away; I need someplace to stay. He's destroying me Christopher. I can't live like this any longer!" she wept with sadden sobs.

"Obviously, Vern doesn't love you-he means to harm you. Pity he still lives with you."

"I'm beginning to wonder where he lives. After I get home, he's out the door all night with his mobster buddies."

"What does he do when he goes out with his hooligans?"

"He gets into fights, he drinks and smokes Meth, and even robs places."

"Are you sure?"

"Yeah."

"How do you know, Darlene?"

"He tells me. I've seen the money he brings home and he's always carrying a gun."

"Do you live in a house?"

"No, we live in an apartment," she said regaining her composure.

"How far away?"

"About a mile or two."

"What floor do you live on?" I asked.

"The bottom floor."

"Good . . . that's good. Do you have any children?"

"No; does it matter?"

"Of course not, but it simplifies things," I said.

"Why all these questions, Christopher?"

"Because I've decided I want to help you, Darlene," I replied with the gentlest smile I could muster. "I have an idea, but in order for me to help, you must do exactly what I say; okay?"

"I'll try anything," she answered.

"All right . . . first, we're going to get a taxi and take you home. When we get there, you'll wait in the cab while I go to the door. If he's there . . ."

"What are you going to do if he is?" she asked worriedly.

"I'll just tell him I'm sorry that I have the wrong place, and then turn around and walk away. After that, then we'll simply wait until he leaves."

"And if he isn't there?" she asked.

"Do you have a key?"

"Yes."

"Good. If he's not there, we'll just go in and pack a suitcase."

"Then what?"

"We'll go to the *West Bridge Hotel* and get you a room for a week. And it'll be important that you don't tell anyone where you're staying-I'll pay. Are you still with me on this?"

"Yes, but do I go back to work?" she asked.

"Hum . . . is your boss a woman or a man?"

"A man."

"Damn."

"What's the matter?"

"I think a woman would understand . . ."

"Oh no . . . he's very nice and understanding."

"Well, good. Call work and tell your boss you have a very serious problem. Explain everything. Ask him if you can take a week off until your bruises start to heal.

If he's as reasonable as you say, I'm sure he'll understand. And don't tell him where you're staying. But do tell him to be careful about Vern. This is important, do you understand why?"

"I think so," she said, trying to understand why.

"Okay, Darlene, I'm not trying to make this sound melodramatic or anything, but I'm truly concerned for your safety: are we clear on everything so far?"

"Are you sure you know what you're doing?" she asked.

"I'm not a hundred percent, but you asked for my help and this is what I've decided-can you think of a better idea?"

"No. Not really."

"Okay, let me finish. After we check you in the hotel, I'll want you to take a taxi to the West Precinct Police and get a restraining order and fill out a police report against your husband. The police will take pictures of your face. After you do that, go to the hospital and have yourself treated, and be sure to get a statement of your injuries. Then go directly back to the hotel, get something to eat and try to get a good night sleep-I'll see you tomorrow, okay?"

"And I suppose you wouldn't what me leaving the hotel?"

"That's right. I'll call and visit you, okay?"

"All right, Christopher, I understand; but you'll come and see me tomorrow won't you?" Darlene asked with a little apprehension, mixed with a fondness in her voice that affected me.

After everything was said, I got up and nonchalantly grabbed my money, and my gun from the bedside drawer. Darlene didn't see. We took a cab to her place, and with astounding luck, or maybe someone above, Vern was gone and everything happened just as I'd planned.

It was about eleven-thirty before I was back in my apartment. I grabbed a cold one from the refrigerator and a bottle from the bathroom. I sat down at the table and poured some scotch in Darlene's empty glass and popped the beer. I took a swig of scotch, a gulp of beer and a drag from a cigarette I just lit. My thoughts were about Darlene's well being, and especially her face. Today, she looked nothing like the day I first met her. My thoughts drifted back in time. Darlene has the most gorgeous green eyes, silky blonde hair and a very soft and supple face. And she has a very nice figure; she doesn't look a day over twenty-five. She had indeed, struck a new desire; a long emotion which I thought I'd lost along time ago.

When we arrived at the hotel, I paid for her room in advance, which took a little over $875 from my pocket. I make a mental calculation to estimate what I had left in my pocket. The taxi was about $35, and the extra money I gave her for food and other necessities came to $250. That left me with about $25 to my name. I wondered what she thought of me.

I couldn't decide if she considered me a fool, or just a dumb ass trying to be cool. Hopefully, I thought, she thinks I'm a person who cares, and is someone she can trust. Only a fool would spend that much money to look cool. I had to make myself stop thinking of her. I continued to drink while playing with the jigsaw puzzle. It suddenly occurred, and surprised me, that Darlene, nor myself, was ever aware when I pushed the puzzle slightly aside. Strange.

I spent the rest of the night, in a relaxed repose, until I knew it was time for bed again.

Boom! Boom! Boom! "Shut the hell up in there! Stop that fucking yelling and screaming . . . or I'm calling the goddamn police!"

Along with my own screaming, and the pounding from the wall, and the people shouting outside the door was all it took to frighten Mr. Sandman awake. Goddamn my life. I got up and went into my nightmare morning mode. It was eleven-thirty in the morning. Instead of going for a beer, I decided a cup of coffee was in order for today.

Still half asleep, I boiled some water and filled my cup with old instant coffee. I returned to the kitchen for some milk and chased the cockroach's away-nasty little things. They can change and evolve just as fast as their environment. I sat at the table sipping my coffee and puffing away at a cigarette.

It was a sunny day, which gave a false impression that it was warm outside. But, spring was just around the corner and I couldn't wait for it to expose itself. I yearn for the colorful flowers, the leaves and grass to be green again.

The winter steals all the colors and the warmth leaving only the stench of cold, whereas spring brings the colors back and breathes a new and refreshed life and the smell of hope. And that feeling feels good. I look forward to this season every year.

Darlene was on my mind again but in a strange and nice way. She gave me something new that I can't seem to describe-maybe something new to look forward too, perhaps even a new future. She was like the season of spring. Thinking about Darlene made the minutes go by quickly, as did my evil dream. I started to feel better and normal again.

The sweating and shaking had passed and I knew I was on my way to recovery. I needed to get cleaned up soon because I wanted to see how Darlene was doing. I also thought about asking Billy if I might borrow some money. It wasn't long before I was out the door and on my way.

Before I went to the hotel, I stopped at a restaurant and ate a $4.95 egg. That was the only thing I ate. This place was too rich, at the moment, for my blood. They must've mistaken me for a rich man, maybe because I was wearing one of my new outfits.

I looked like a million dollars, but in reality, I only had twenty bucks on me. When I reached the hotel, I went to the lobby desk and asked them to call Darlene.

"I hope I haven't disturbed you madam, but there is a gentleman by the name of, Mr. Christopher Lance, who wishes to see you-may I send him up?" I waited with anticipation.

"She's in room 910, sir."

"Thank you very much, young lady," I said to the desk clerk like a gentleman.

I knocked on Darlene's door and in a few seconds, I heard a latch click and watched as the door open. She looked like hell.

"How are you doing? Are you okay?" I asked.

"Yes, but my face hurts so bad."

"Did the doctors at the hospital give you any pain medication?"

"Yes, I just took a *Vicodin* tablet and some special eye drops. The phone woke me up when the desk clerk called," she said yawning.

"I'm sorry, Darlene, I didn't think you would still be asleep. Please, go back and lie down-get some more sleep."

"Yes, Christopher, I think I will, but please don't forget me. Come back and see me again later on."

"I've got your phone number, Darlene. I'll call you, and don't accept any visits or calls unless you know it's me . . . all right?"

She nodded yes and slowly wobbled back to her bed."

I grabbed her room cardkey and quietly left. I decided not to bother Billy with money and headed for home. I felt sad. On my way home, I stopped at Darlene's apartment, walked inside, and casually walked around. I noticed a broken window but kept walking, while wishing subconsciously that I'd run into Vern. The place was a mess. I saw furniture tipped over and broken glass on the floor. There were holes in the walls and pictures thrown about, and the TV's picture tube was smashed in. All in all, I'd say that Vern made a picture prefect mess. Apparently, Mr. Vern Osmon received his restraining order. He must've come back for revenge. I thanked God that Darlene was safe. As far as Vern knew, she had simply vanished from the face of the earth. Good riddance to you, I thought, you fucking woman beater.

I headed back home and changed my clothes. Again, my new clothes went back into their respective plastic covers. I put on my usual get-around garb, and sat at the table feeling rather confident and successful for a change. I looked at my watch, it was seven o'clock, and in an hour or less, it would be dark.

I grabbed my drinks and sat at the table staring blankly out the window. I suppose I decided not to go and ask Billy for any money because I knew he really couldn't afford to loan me any real sum of money anyway.

I sat and drank and smoked, while trying to figure out how I could come up with some money. It was actually obvious to me, but I had to coerce myself to even think about it again. I've done this before, but only once, and I considered myself lucky to get out alive. It's an extremely dangerous thing to do, but I told myself that I must. It was about nine and dark outside. Stealing money from a dope house was a very risky business. I liken it to a suicide run. Russian roulette, if you will. I knew the house I wanted to hit, and then walked over to my magic bag of tricks. I needed to look like a junkie for this little gig. I first applied my face with makeup. I used a small amount of white toner and smeared it all over my face, rubbing it in good until I had a pale and tired face. I then put some blue and red masque under my eyes, trying for a slight look of sickness.

After the makeup, I put on some scabby clothes and a short ratty wig, which finished the look. I took my last $20 and rolled it up with some fake bills under, and tied the wad with a rubber band. An old trick, but many times it still works. The last thing I did was dilute some water with cleaning solution containing a very small portion of chlorine, and splashed it in my eyes. It burned slightly for a few seconds, but just like a swimming pool, it made my eyes blood-shot-red, but not so bad as to impair my vision.

I put on an old, flimsy gray coat that had a few holes and looked like a rag. I filled all the chambers in my gun, and slide it in my right pocket. Then I took the stairwell steps down and went out a back door. The last thing I needed was a smell.

I rubbed my hands in some dirt and then on my clothes, with just a little on my face. And for the final touch, I crawled into a dumpster and smelled myself up.

I went back into my apartment to have one final look. I smelled bad, looked bad, and wore an expression that junkies have. It was a face that was hard to remember. I took another generous drink, and then walked out the door to pursue my prey.

CHAPTER FOUR

The drug house was about six blocks south of Steward Street, where the neighborhood is even worse. I walked the distance ducking down alleys and side streets to avoid other people. I finally reached the house and quietly surveyed the parameter-first at a distance and then up close. There was a boarded-up house on one side, and on the other, a small weedy lot that bared the scar of an old unfinished foundation.

I saw some evidence of life in the house as I walked closer. I was now spying in the windows, while taking a few quick glances behind me. I wanted a surveillance of 360 degrees around me. As I was stalking their crack house, I felt like a beast stalking its kill.

I looked in one window and saw someone lying on a flat, nasty-brown mattress. He looked dead to me-maybe an O.D.

His eyelids were wide open but his eyeballs were rolled back in his head, giving his eyes the appearance of white death. I continued around the house and saw some children sleeping in another room.

"Damn-it-all!" I whispered to myself, "What the hell are children doing here! Where are their mothers?"

I concluded that their mother's were probably prostitutes; working the streets while their boyfriends, or whatever they were, sitting at *home* on their lazy asses selling dope. I almost abandoned my assault but decided to continue, assuring myself that the police will take care of the kids.

I went to another window and saw nothing. I then peered through another window, which turned out to be the kitchen. I found myself staring at a filthy sink, and smelling spoiled food through a split in the bottom ledge of the window. But again, I saw no one. Finally, I came around to the living room and saw three guys sitting like zombies and listening to some really funky and ugly music. One guy sat in an old, crooked armchair with stuffing falling out, and the other two where setting on an un-cushioned and sagging couch. In front of them sat a heavily stained and scratched coffee table. And on that table sat the booty: money, coke, crack, pills, weed, and pipes-the whole damn rotten mess.

I made one last tiptoed trip around the house making sure I hadn't missed anything. In the little time I used encircling my prey, I saw no one come or go. After I was satisfied with my surveillance, it was now time to make my appearance.

I took a deep breath to gain my composure and went into action. I half-stumbled and half-walked up the broken steps and knocked on an old flaking door. Gazing cautiously through a small front window, I saw a guy sitting on the couch get up to open the door ajar.

"Yeah?" he said with baited breath.

"I heard that I could buy some sugar here," I said, stressing my face and slurring my words.

"Well mister, you heard wrong," he muddled back and proceed to close the door.

I jammed the door open with my foot, "come on homies, I need a fix and you're my last chance. Can't you give a man just a little piece of gratification? I have money."

"Move your foot out of my goddamn doorway, or I swear my face will be the last thing you'll ever see," he said with conviction.

"Okay, okay, but let me show you something. Look at this," I garbed my fake wad of $20s, waving it back and forth.

"Jake told me I could find some sugar here-remember Jake? He told me you guys partied together . . . about a week ago?" I lied to the drug dealer, still keeping my wits about me, and yet still performing my act as a desperate junkie.

"Hey, you guys?" the dealer asked the others without turning his head, "Any one here know a Jake?"

"I remember him," the guy in the armchair grumbled in a deep and rash tone over the music.

"Yeah, we know Jake. You said you have money?"

"Yeah . . . it's right here in my hand-see?" I said showing the fake wad of cash.

"Come on in," the guy on the couch growled, as the other guy opened the door. I faked a small stumble and fell on one knee as I entered. I got back up and waved the fake wad of $20s around.

"I've got $200 here so I can snort some coke," I mumbled loudly, and then retched and gagged a little; but I was not pretending.

That little show was for real. The smell of that place was despicable; a truly unique odor-I felt sorry for the children sleeping in the other room.

"What's your name?" the guy on the couch asked.

"My name's Willy Ashburn," I muttered, when all the while I was scheming and sizing up the situation, and calculating my next move.

"So you want $200 worth of cocaine?" the guy in the armchair said. The three starting laughing when I humored them by doing a little dance in a circle. At a very sudden moment, and in a split second, I had my gun out and trained on the man in the armchair-the silencer showing.

"Okay guys, I think you all know what to do. Empty all your pockets and put your money and guns on the table-all of it. And if I see any sudden motion, I think you know what I'll do next." I demanded.

I knew that they were all right handed from my observations. The guy in the armchair made a quick jerk, but before he had time to pull his right hand away, I had two holes in his shoulder. Faster then an eye blink, I had my gun trained steadily on the other two. The guy in the armchair was moaning.

"Now . . . you two!" I barked, "Move your hands where I can see them, and slowly put all your money on the table; Now! Let's move!"

They unhurriedly reached in their pockets and placed their money on the table. "You too, mister!" I said to the guy with the gruffly voice and bloody shoulder. There was a substantial amount of money being placed on the table; but I knew it to be better to keep my eyes on these clowns instead of the money.

"Okay, you. Yeah you!" I gestured with the barrel of my gun, "Put all that cash in the gym bag lying next to the chair-now!" I shouted, and then shot the music machine to destroy that God-awful noise. The sparks and the sounds of the plastic shrapnel bouncing made more noise than the silencer on my gun.

"Zip the bag up!" I shouted again, and he closed the gym bag just like I told him. All three of them looked at me like mad dogs. With blood in their eyes and snot in the noses, they snorted and mouthed obscenities bubbling from the foam on the corner of their mouths.

In a blink of an eye, I shot the two guys on the couch each in their crotch. I turned quickly and shot the guy in the armchair in the crotch too. They all started kicking and screaming. I was now in the process of letting the distraction and shock give me enough time to quickly slip away.

I grabbed the gym bag and hurried out the door and down the streets. I calmly walked away quickly, but not as to draw any attention. I just continued through the side streets and alley-ways-zigzagging my way home.

Inside my stairwell, I ran up the steps like an athlete training for the Olympics. When I made it back to my apartment and threw the bag on my bed, I sat down for a while just to catch my breath. My heart was pounding and my breathing was heavy. Only now did my hands start shaking when I knew I make a good steal, but that never stopped me from counting my money before.

"Holy shit!" I yelled, and then whispered, "$10,855."

I put the money in my drawer, put my wig away and took a shower. After I was cleaned, I put on a pair of pants and a T-shirt to wear. I grabbed a cold beer and slugged it down cheerfully. I retrieved another cold one and sat at the table. I was starting to calm down. I poured a glass of scotch and lit a cigarette. I told myself that I'd never, ever do that again. Looking at my watch, only a little over two hours had passed-but it felt more like four of five. I collapsed on my bed and fell asleep on my back.

"Oh don't. No please. Stop them . . . don't let them do that like . . . God, stop them!"

Another damned night, good old Sandman sometimes sends me deeper and deeper into these goddamned dreams. The eye dream-that one really fucks me up. It has the most tormenting visions and painful feelings of all of my nightmares; it's one of my cruelest and most unthinkable night terrors I have. I can hear, feel and even taste every little nasty slice of that dream-and the pain is fiendish. Thank God it was only a nightmare. I wanted a drink, but decided instead, I'd better have coffee. It was ten forty-five and there was still a whole day ahead of me.

I wanted to see Darlene again. She was the only nice thing that felt good in my life. After a few drinks of coffee, I felt better. I already took my aspirins and the sweats and shaking were going away. I lit a cigarette and pulled the curtain open for a look outside. The sidewalks were busy with working people who make the world a better place. Not like the night-street people, who lay waste and cause pain, and then vanish from the dawn of the day. I realize I'm one of them, but my life differs in many ways. As a matter of fact, my very existence sets me apart. And I am a survivor-young enough to live through it. I don't kill or sell drugs to meet my needs-I just risk night walks.

But, last night . . . last night was one of the most foolish things anyone could do. Even the police hate to bust drug homes; and they even have backup. Someone could have walked in while I was robbing them, or maybe I could have been shot in the back. It's all in the game.

Very few people in my situation would not have done, or be bold enough, to do what I did last night. But I must be honest with myself; the more I do this kind of shit, the percentage of me dying increases with each incident. Eventually, someone will kill me. I told myself last night that I had to stop this crap. Just because I've never been caught doesn't mean that the next time I go out, I wouldn't be the one who gets shot-probably just the opposite.

My thoughts switched back to Darlene and my desire to see her again. After a few hours, I was out the door all dressed up in a different outfit. Before leaving the apartment, I took $1,500 with me and stopped at that store where I had bought my other new clothes. I bought two new pair of pants, two sport coats and shirts to match. I also included a new pair of shoes just like my black ones except these were brown. I left my merchandise to pick up later. About three blocks before the hotel, I stopped at a gun shop and purchased a shoulder holster. I wore it out of the store underneath my sport coat. Of course, my gun was in the drawer back at my place, but I decided I was no longer going to carry it in my pocket. A holster makes everything easier, and easier to get my gun. Besides, it was the professional thing to do.

It was about three when I knocked on Darlene's door. "Who is it?" I heard her voice say through an intercom that I hadn't noticed before.

"It's Christopher, Darlene," I replied with a push of a button, and then the door slowly opened. "How are you feeling?"

"A little better I think. My face still hurts, but it doesn't feel as swollen. The doctors told me that nothing was broken and that it would just be a matter of time before the swelling goes away. He gave me some prescriptions to take, and as you can probably tell-I'm doing fine. So that's my situation, how are you doing today?"

"I'm okay. Did you call your boss?"

"Yes. He was so nice and understanding. He told me to take a week off, or if necessary, two so that I could get my life back in order. I told him everything about Vern, and he completely understood. I told him I'd call back in about five days to check in."

"Did you tell him where you were calling from?"

"No-but he didn't ask anyway. Nobody knows where I am or my phone number except you, Christopher."

"Good."

"So what's the game plan now?" she asked.

"You're going to stay here for awhile; unless you want to see Vern?"

"No fucking way! Oh . . . I'm sorry I swore; it slipped out."

"Words don't bother me Darlene, say whatever you want. You were saying . . ."

"I'm going to file for divorce. This ends it."

"How many times has Vern beaten you?"

"I'm not sure-maybe five or six. Before we were married, he never gave me any signs of being abusive. Now it's come down to this shit. Guess what the police told me?"

"What?"

"They told me that he's wanted on pervious charges-including solicitation for prostitution. I also learned that he's wanted for robbery, and that he's suspected of killing the store clerk. All this shit just happened recently. The police said that they've been looking for him for over two weeks. He's been doing this crap at night-when he goes out with his boys-the lousy bastard."

"Darlene?"

"Yeah?"

"I stopped at your apartment yesterday and everything's torn up. My first guess was that he fled from the police when they caught up with him, and then came back later hoping to find you. But that doesn't explain how he made it in and out of there without the police seeing him. Then I thought, maybe he tore the place apart before the police came, or maybe because you didn't come home?"

"How did you get in my apartment?" she asked.

"I walked in."

"Then why couldn't have he?"

"Good point."

"What did he ruin?"

"Everything."

"How about my computer?"

"I didn't see one."

"Goddamn it; that rotten son of bitch! Oops-I'm sorry-I swore again."

"Don't worry about it," I simply said, "You'll probably never see your computer again."

"He'll sell it on the streets, I'll bet."

"That would be my guess."

"Did he leave anything intact?"

"A few things . . . but I would rule out your television, most of the furniture, even some pictures on the walls."

"Okay, Christopher, I've heard enough-I get the picture."

"I'm sorry, Darlene."

"Don't be. Why should you? Most of it was junk anyway and besides, it would just remind me of him. Fuck it. I'm getting hungry."

"You're hungry?"

"Yes, I'm ravenous."

"Tell you what. How about I order a nice dinner for the two of us. We can eat at the table and look at the park below.

Have a little conversation and enjoy some time together-sound good to you?"

"That's sounds very nice, Christopher. But I need to tell you something."

"What is it?" I asked, feeling unsure what she was about to say.

"I want to be honest with you, I don't know how soon I can pay you back all your money. You've got to be my guardian angel."

I smiled and told her that I was definitely no angel, just a person who decided to help a pretty lady. I didn't know then what I know now and I suppose, neither did she. We had a wonderful time together that afternoon. We ate and talked while we watched the adults and the children, stroll and play in the park.

I told her she was very pretty; I couldn't say it enough, but I also told her she was intelligent and fun to be with. She spoke nice words of me. We seemed to hit it off. But every day has its ending, and this day was no different than the rest. Later in the evening, I told her I should go. But she wanted me to stay a little longer, saying she felt safe with me around. Darlene also told me that at night she gets bored and lonely. I wasn't sure how far this would go, but I stayed a little longer and we talked. She started talking about work and home, and then an unwanted question popped its ugly little head.

"Christopher, you've changed since I first met you, you dress so handsomely; why do you live in that apartment?"

I spit out some bullshit, "I really can't tell you. I'm not sure myself, but for some reason, I like it there. Maybe I'll move somewhere this summer that's better."

"What do you do, Christopher?"

"I'm trying to be a writer," I said, and thought to myself that every damn bum without a job is an aspiring writer. But not me, I'm just a damn bum.

"Hey; maybe you could give me some of your work to look over. Remember, I'm an assistant editor at Martin & Kelly."

"We'll see, Darlene. But I want to be published for my work alone, not because of who I know."

"Oh, don't be so trite. There are gobs of people out there like you who'd give their souls to get published. So if you haven't been published, how do you make your living in between your writing?"

"I just do odd jobs here and there, and sometimes for my friend, Billy. That's one of the reasons I live in that apartment-it's cheap. No overhead."

"But I didn't see any writing equipment. I didn't see a computer or even a typewriter."

"I write long hand." I replied quickly.

"You've got to be kidding. I could teach you how to use a word processor on my computer . . ." her voice slowly faded.

"You mean your stolen computer."

"Yeah, that one. Oh well, my insurance should cover it."

"All right, Darlene, let's get back to you," I said changing the subject. "We've got to get you back on your feet, back to work and living somewhere else. Some place where your husband can't find you; unless you want him back?" I suggested once more.

"Christopher, I told you. I don't want him, I'm getting divorced."

"Okay then, enough of this matter; how are you set with money?"

"Not very well. I have about $150 in my savings account and when I go back to work I'll . . ."

"No no, Darlene. I mean, how much money do you have with you now?"

"About sixty-five dollars."

"Okay. Here," I said digging in my pocket," Take this."

"Five-hundred? What for?"

"Just hang onto it in case you need it. You have to have some cash on hand. There are a couple of clothing stores, a bookstore with magazines and a nice restaurant next to the lobby; use them."

She refused the money, but I'm not stupid. She needed it; she knew it and I knew it. Darlene pretended a little more, and so I played the little game and insisted she take it. She took the money. I would have to question her intelligence if she didn't. I knew that sitting around in a room gets boring. I wanted her to be able to get out for a while and keep herself busy and hopefully, happy. She promised to pay all the money back, but little did she know at the moment, all the money in the world wouldn't make me happy-I was beginning to think I knew what would. Maybe some feelings that just go a little deeper then the words: thank you.

I'd have to say, that I felt a little embarrassed at myself for what I had been thinking. I told her I must go and did, but as I was leaving, she gave me big hug and kissed my cheek, which took away some of my hidden embarrassment.

It was twelve-thirty when I arrived home and saw Mr. Scott Sellers sitting in the hall next to my door. "Hello, stranger," I said with a smile and a quick glance. Scott didn't say anything. I looked back down and saw that he was holding his right leg and bleeding badly. I looked behind me and saw spots of blood which I hadn't noticed before.

"Scott, what's wrong? What's happened?"

"I was stabbed in the leg, Sandman," he said with a special grin on his face-the grin that only shows pain.

I unlocked my door and helped Scott in, and then bent down to look at him. "Let me look at your leg; let go of your leg so I can see how bad it is." There was a deep cut in the shin of his leg. "Put your hands back on top of the cut and apply pressure here while I change." I was wearing my new clothes and decided to quickly change so I wouldn't get any blood on them. As I was changing my clothes, I remembered that I had forgotten to pick up my new clothes I had bought earlier today. I'll get them tomorrow I thought to myself.

"What do you want me to do here-bleed to death?" Scott's voice snapped causing my thoughts to come back to the incident.

"How did you get stabbed?"

"Some street shits wanted to use me as entertainment. You know what I mean. Kick me, slap me around, and made me dance and sing. They apparently were dissatisfied with my clowning around and two of them pushed me down. Then one of the guys drew a knife and I kicked my leg at him. Then I felt the pain and that's when they ran away."

"What did these punks look like?"

"I don't know, Sandman, just your typical little gang-just kids," Scott replied with a moan.

"It's a pretty deep gash, would you like to go to the ER?" I asked.

"No," he proclaimed.

The streets are never safe at night. These stupid street gangs like to terrorize innocent people, and then rob and beat them to death. They're savages. The little cocksuckers haven't a clue about the immeasurable price of life. But how could they, they're not old enough yet? They really pissed me off; I'd like to go out there and corkscrew one of those little sons of bitches right in the ass with a serrated, ten-inch knife.

"Okay, Scott, I want you to stick your leg in the shower. Be sure that you wash it with soap. Let your cut bleed a little more and when you're through, I'll take care of your cut."

I heard him bitching and swearing and moaning in pain. While Scott was cleaning his leg, I was already preparing some of the necessary items I'd need. After a couple of minutes, he came out dragging his leg with a towel wrapped around it. His blood was seeping through.

"Okay, Scott, this might sting a little bit, but I'm going to pour some hydrogen peroxide and iodine on it. Put your foot in this pan." I poured some peroxide in is wound, and then some iodine.

"You're wrong, Sandman, it doesn't sting; it burns like hell," he said ignoring the pain.

"Well, the worst is over. Now, hold this cloth over your cut for a minute-and keep the pressure on."

As he was doing that, I was preparing the needle for stitches.

"What do you think you're doing?" he said.

"It's a deep and lengthy cut. I'm going to stitch you up-six ought to be enough."

"Have you done this before?" he asked.

"Twice. Okay, Scott, remove the cloth from your leg."

I sewed the stitches spacing them closer where the cut was deepest. Scott never uttered a sound, but watched me closely with that pain grin of his. After the stitches were in, the heavy bleeding stopped. I applied a little bit of antibiotic salve on it, and then placed sterile gauze on top securing it with medical tape.

"You're going to have to stay off that leg for a day or two, Scott. I'm afraid if you walk on it and twist it, those stitches will come lose and your leg will reopen and start bleeding again-and it could possibly get infected. That's a serious deep cut and it's nothing to take lightly."

"Okay, Sandman, but where will I go, where will I stay?"

"You'll stay here with me. It's a damn good thing I'm your friend or I would've let you bleed to death out in the hallway. Sorry Scott, but that's the way I am."

"You know, Sandman, I think you're a miracle. You truly are an amazing man."

"Where the hell did that come from? I don't know what you're talking about. You're not making any sense, Scott. Don't say things like that.

I'm not a miracle, I'm just a person like everyone else . . . see?" I said with my hands in the air.

"Okay, Sandman, but I don't care what you say, I think your world is going to change in a very unique way."

"Yeah? Maybe I'll find a stash of cash," I said and smiled, "How about a drink?"

"Sounds good to me. I've been sober for three days," Scott answered.

"Well, maybe you shouldn't be drinking. You already have a three day head start."

"Fuck sobriety. You still got any of that scotch, Sandman?"

"Yes. Are you really sure you want some?"

"Sure as I know I'm going to die someday," he replied.

I took two clean glasses from the cupboard and two cold beers from the refrigerator. From the bathroom came a new bottle of scotch and a couple of packs of cigarettes. We sat down, popped our beers, poured our scotch and lit cigarettes. Then we started a bullshit session.

"How's your leg feeling now, Scott?"

"It's starting to ache, but I'll be okay, son."

"What did you call me, Scott?"

"Oh. I think I called you . . . son. I'm sorry Sandman, I wasn't thinking."

"No, no. That's okay; don't be sorry. That's the first time I've ever heard anybody call me that. Sounds pretty nice actually."

"What do you mean, Sandman? I know you have a father and a mother."

"I have a father, but no mother-I've never seen him in my entire life."

"You can't be serious, Sandman?"

"I'm as serious as a deadly snake bite," I said with a grin and a blink of an eye.

"Where were you born, Sandman, where did you come from?"

"I come from a very strange place, Scott, and I don't think you could, or would understand. I come from a different place and another time."

"I don't follow you, Sandman; are you trying to tell me you're some kind of alien from outer space. You're not making any sense; what are you talking about?"

"Oh, I'm just bullshitting you; just gibberish, Scott. I'm making fun."

"Strange way of joking," Scott mumbled.

I decided to change the subject. "I'll have to change your bandages tomorrow, but I know you're going to have to stay here for a couple of days until that wound starts to heal."

"That's very kind of you, Sandman, but I really don't want to take advantage of your hospitality."

"Listen here, Scott, nobody takes advantage of me, and you're not imposing. I just don't want you running around the streets with your leg in that condition. You wouldn't last two days out there."

"Thanks for letting me stay, Sandman-you know what?"

"What?" I echoed.

"There's something about you that's very unexplainable. Something inside you that emanates: assurance, safety, confidence, camaraderie and most important of all—friendship. I've never seen all those qualities in one person before."

"You're babbling, Scott, let me refill your drink and get two more cold ones."

"What are you doing with all the puzzle pieces on the table?"

"Oh, I thought it would take my mind off the nights. It seemed like it would be kind of fun at the time."

I picked up the puzzle pieces and put them back into the box and sat them on the kitchen counter. I looked at my watch-one in the morning. I returned to the table, turned on some nice music and kept the volume low for some background melody.

"I saw you leaving the parking lot the other day, Scott."

"You did. Well, it beats sleeping in a cardboard box."

"How much do you pay the security guy for one of those back-seat hotels," I asked.

"Five bucks a night."

"Hum, not too bad," I answered. "What do you do during the daylight hours besides begging for money?"

"I study."

"What do you mean, you study."

"Well, I go to the library and study books."

"Library, huh. That's good, Scott, that's very nice. I'm proud of you. You're going back for your high school diploma-is that it?"

"No, Sandman. I'm beyond that."

"You mean to tell me that you already have your high school diploma?"

"That's right. I took a high school equivalence test and they gave me a GED."

"Scott, why didn't you tell me this before?"

"I didn't think it was any big deal."

"Scott, this is great! You're not lying to me are you?"

"No."

"What kind of books are you reading?" I asked excitedly.

We both took another few shots of scotch from our glasses and sucked down more beer. I lit another cigarette and Scott did the same.

"Well, basically science and math. I've just finished an advanced book on differential equations and now I'm studying differential geometry and tensor analysis."

"You're talking about math, aren't you? I know it can be a very deep and intensive field of study. I also have an interest in numerals and numbers. Tell me, Scott, what other subjects have you studied?"

"I've studied everything from A to Z and any subject that you want to ask me, I'll give you the answer," he slurred intelligently.

"Damn, Scott-you must be a genius. Do you really understand all that advanced shit?"

"Yes, and I have perfect recall too."

"You can honestly sit here and tell me that you can recall everything you've read?"

"Oh, yes, and I have total memory recall, including graphs and pictures."

I found myself drinking more then Scott, and I was also chain-smoking. Here I sit, with someone whom I believed was half-witted, and now, this someone suddenly blows my mind into scattered little fragments. His advance knowledge in the sciences appeared to be incomprehensible; he could explain any complex subject in layman terms.

He then answered my questions to which I had no answer. I just sat at the table thunderstruck.

"Sandman, I'm really nothing more than a bohemian scholar. I started going to the library one day to get out of the cold and picked up a book on art. I became interested and started to make it a practice to read. I already told you that my dad wouldn't let my go to school, so after I was laid off from work, I decided to educate myself.

I've been studying on my own for over thirty-eight years. I've progressively learned to understand one book after another until I was ready for the next. Many times I've studied three or four subjects at once. As I continued this routine, I began to understand that I was capable of retaining all of this knowledge too. But I'll remind you again, this has been over a span of thirty-eight years."

After awhile it suddenly occurred to me. "Scott, if you have all this knowledge, including your aged wisdom, why in the hell are you a fucking bum?"

"Oh, that's easy. I love knowledge for the sake of knowledge itself-it brings to me a feeling of magic in my soul. Although it's true I will not use all of my knowledge, there's something kind of nice just knowing and understanding these strange and complex ideas written in abstract language. I absolutely love it!"

"Then what you're telling me is; you're not interested in applying any of this stuff to creative and productive work?"

"That's exactly what I'm telling you."

"But why, Scott?"

"I already told you."

"Yes, yes, a deep understanding . . . but you could have any job you want. You could be a . . . mathematician, a chemist, a rocket scientist, a computer engineer, a teacher, a mechanical . . ."

"I don't want to be any of those things, Sandman."

I took a big gulp of scotch and then another. I paused for a minute to light a cigarette and gathered my thoughts while I tried to make sense of all the shit he was talking about.

"Okay, Scott, then why all this crap about *acting* like a fucking bum. Are you trying to fuck with me? Take my money? I sewed your damn leg up. You've slept here, and drank here. You've eaten here and even shit here! Why, Scott, why are you playing me for a fool? Tell my why, Scott . . . why?"

And with all his knowledge and wisdom his singular reply was, "Friendship."

"Friendship! Is that the only damn excuse you can come up with from that fantastic mind of yours?" I yelled in a semi-drunken stupor.

In a gentle, calm and soft spoken voice, he answered, "Please remember, Mr. Sandman, it was you who sought me out. It was you who wanted to know what I was all about. It was you who wanted to know my ugly truth. I didn't withhold any information about my journey through my youth. However, I did not tell you all of the truth because I wanted you to know me for who I am, not for what I know. I didn't want to lose my friendship with you either-it means a lot to me. I swear on my mother's grave, Mr. Sandman, I haven't had anybody call me friend for years before I met you. I'm beginning to feel a likeness toward you. Don't misunderstand me here, I mean as a brother, likeness as a father and likeness as a friend. I need you Mr. Sandman, more than you know.

CHAPTER FIVE

All the knowledge in the world means nothing, if one doesn't have at least one true friendship. I'm pleading with you-please, Mr. Sandman, be my friend. You already said I could stay."

"Oh, quit babbling like a fuckin' beggar, I'm not going to throw you out Scott. You're too close to me now-like a damn puppy dog. But I am happy that I found your friendship."

"Thanks, Sandman. Then we're still friends?" he asked.

"Friends like a father and son," I answered him with a big, stupid drunken grin.

It was a little after three in the morning and we decided to celebrate for a few more hours. We took turns looking out the window and at each other as we talked. We even toasted ourselves to our newfound friendship. We were drunk, but far from passing out. I was still amazed at the amount of knowledge he had stored in that mind of his. He was the most intelligent man I ever met, and at the moment, I was proud to call him my friend. What he saw in me, I'll never know, but the important thing was that he liked me-I was his friend too. We drank, smoked and talked some more. I couldn't help but ask him some more questions that I alone could never figure out. I took a drink of scotch, a slug of beer and a drag on my cigarette before I got the nerve up to asked him a trick question I learned from a magazine.

"Okay, Scott, here's one. If it's zero degrees cold outside and the weather person says it's going to be twice as cold tomorrow, how cold will it be?"

"Are you serious, or are you making fun of me?" Scott said with a crooked grin.

"Seriously, Scott." I said while trying to keep a straight drunken face. "The temperature will be -32 degrees if you're talking Fahrenheit."

"Hey! You're right; I should have known better then to try and fool you."

"I don't mind trick questions, they keep me on my toes," Scott replied.

"Okay, Scott, here's another one, a serious one though-something I don't understand—what is fire?"

"Ions. You've heard of them, haven't you?"

"Hum . . . I know they have something to do with atoms, but I really don't know what they are."

"Fire is the fourth state of matter, Sandman. A phase of matter called plasma. Plasma is composed of ions, which are made of atoms and molecules that have lost or gained one or more electrons. When this happens, they become electrically charged-both positive and negative-and then they're called atomic and molecular ions. In the case of a fire flame, when a combustible material is subjected to heat in the presence of oxygen, you create a maelstrom of these little things. This is what you're looking at when you look at a flame."

"So the flame in a fire is a bunch of ions?"

"You got it. And they're hot like buggers; aren't they?"

"Is this a nuclear reaction?" I asked.

"No. It's a chemical reaction. Nuclear reactions are an entirely different phenomenon."

"And what phase of matter is that?"

"It's not a phase of matter-it's more a state of energy; something I'd rather not get into tonight. But now you know what fire is, I hope."

"Well, sort of; but you'll have to draw me some pictures," I asked, but he never complied.

We talked and drank some more, and the time went by so fast. I know the both of us were getting tired and drunk. But we were having fun, and we also had established a newfound bond between us. Maybe something special, I really didn't know. I looked at my watch and it read five o'clock in the morning.

"Are you getting tired, Scott?"

"Not really. I'm positively enjoying your companionship, my friend."

"Then to hell with sleep. We'll have fun tonight and suffer the consequences tomorrow. A toast to us, Scott," I said getting a second wind.

We held our glasses in our outstretched arms and clinked them together. "To a long and happy friendship, and may God have mercy on our souls," I toasted. We finished our glasses and refilled them, and then I retrieved another two beers. I lit another cigarette, and again, Scott did the same.

"Scott?"

"What?"

"You know what I really like about you now?" I slurred.

"What's that, Sandman?"

"Not only are you a friend, but you're also a fuckin' genius!" I said and laughed out loud, then continued, "You have answers to questions that I have pondered all my life. Can I ask you just one more question, Scott, and then I promise not to ask any more tonight-okay?"

"Go ahead, Sandman, ask away."

"Why do magnets attract and repel?"

"You might be too drunk for the answer? Scott told me.

"Oh bullshit, I can drink you under the table, my friend," I babbled.

"Let's save that one for another time, Sandman, okay?"

"Okay, buddy-old-boy."

"Now, may I ask *you* a question, my dear friend, Mr. Sandman?"

"You just did," I said laughing, "okay . . . okay, shoot away," I giggled and took another drink.

"Why do you keep a gun in your pocket?"

Suddenly, I sobered up a little. "That's an easy question to answer. For protection."

"Protection from what?"

"From the many desperate and destitute people on the streets-especially at night."

"Do you ever carry your gun in the daytime?"

"Sometimes."

"Have you ever shot anyone, Sandman?"

"Yes."

"Have you ever killed anyone?"

"Not that I know of, Scott, I think I've been lucky about that."

"Why do you have a silencer on your gun?"

"Come on, Scott, I thought you were smart. So I don't create a disturbance. Look at you right now-look at your wounded leg. That could've happened to me. But I don't think it would have-that's why I carry a gun. Does that satisfy your question?"

"I suppose it'll have to do, Sandman," Scott said without blinking.

I wasn't sure if I had really satisfied his curiosity, but I dropped the subject immediately. "Scott, I'm drunk and tired. I think I'll go to bed. I have some things to do later today; I hope you don't mind?" But, Scott agreed that he too, wanted to get some sleep, and so I made a bed on the floor while Scott slept in mine. At least this time my bed was clean, and so was he.

"Mr. Sandman, wake up. Wake up!"

Scott had chased the evil Sandman away. I was sweating, shaking, and had a terrible headache. I also had an unusual pain on the back of my neck like I was stung by a wasp. I grabbed my notebook and wrote down some awful information about that damned eye dream again. God, please help me I said to myself as I scribbled vile words in my notebook. It was three-thirty in the afternoon. After about an hour, I had adjusted to a state of partial normalcy. I had to force myself to get off my ass and take a shower with the determination of approaching the rest of the day with a clean and sober brain.

As I left the apartment, I told Scott to stay put and headed for the clothes store to pick up my new outfits. On my way back home, I called Darlene and told her that I wouldn't be able to see her today, but I promised I'd visit tomorrow. She was disappointed until I told her that we would look for a new place for her to live. We talked on the phone for a little while, and then we said our good-byes.

When I reached my apartment, I hung up my new clothes and discovered that Scott was taking a shower. I felt a little exhausted and tired from last night's drinking, and of course, my nightmare. I fixed myself a cup of coffee and sat down at the table with my head hanging low and running my fingers through my hair. Between Darlene, Scott, maybe Billy and even myself, I was starting to become confused and uncertain of what the future would bring.

However, my exposure to the day outside did tell me one nice thing; it was finally turning to spring. I sat at the table sipping coffee, smoking, and waiting for Scott. My thoughts started to turn to Scott. I knew I had to re-dress his wounded leg, and later eat. As I started to look around,

I noticed that Scott had made the bed and picked up the trash from yesterday. I got up and knocked on the wall beside the bathroom door.

"Scott, are you in there?"

"There's nobody in the bathroom right now, Sandman, try a little later," came a reply.

I thought to myself . . . ask a stupid question? It was five-thirty and for the most part, my chores were done except for Scott. I looked out the window and watched the busy sidewalk with all the people going home from their jobs. I saw a person who looked out of place. I then realized that he was not coming home from work, but instead, going. He was a clever little chap-he was a pocket-picker. People, mostly men, hadn't the slightest idea that they were being robbed. With a gentle slip of his hand, he would extract a wallet from an unsuspecting man and put it into a bag. A thought then occurred to me: a Bandit-Yahoo-a robber robbing another robber. Only this time, the robber who would rob the robber, would be me. But not only that, I knew he had money, and it would be a simple little game to play with this man before I steal his stolen money.

I went downstairs and trained my eyes on the thief until he decided he had had enough. When he'd figured he was done for the day, he would then become my prey. Even the predators have their prey, and besides, when I rob the little dude, what's he going to do-call the police? I told Scott that I'd be back in an hour or two . . . maybe three.

I went down to Browne Street, crossed at the light and observed the pickpocket thief from the other side of the street. I didn't let him out of my sight; he will indeed, become his own victim in the end-and I'll win. Little did he know that his dirty deeds would become my windfall. I walked up and down the other side of the street, watching every move he made. This went on for most of day, and I was beginning to wonder how much longer this shit would go on, and how much money he had stolen.

I looked at my watch-it was seven and the sun was beginning to set, turning the day to dusk. By now, most of the working people were at home watching their favorite television shows and eating, or drinking, or fucking, fighting, and all the other weird shit that *normal* people do.

Finally, the pocket-picker started heading toward east; down Steward Street and to the worst side of the city. I crossed the street and followed him until it was almost dark. I stayed close to him while trying to be as stealthy as possible. I ducked into doorways; hid behind trees and bushes, and everything else I could find so as not to draw his attention. The neighborhood was getting bad, and I knew it was time to make my move before I got too deep in the delta. He suddenly turned down a side street. I hurried a bit and turned the corner. There he was, standing about ten feet away, pointed a gun in my direction.

"Looking for someone?" he said.

"Well, as a matter of fact, I am," I said plain and simple.

"Why are you following me?" he asked in a sour note.

"Well, to put it bluntly, I'm gonna steal your money."

"You do realize that I have a gun pointed at your head, or are you just stupid?"

"No. I'm not stupid, but if you don't put that gun down, I'll be forced to take it away, along with your livelihood," I calmly said.

"And who the hell do you think you are?"

"Your thief. Even thief's have their thieves. Now don't tell me you've never heard the old phrase: thieves among thieves. Welcome to your . . ."

While I was still talking, I surprised him with my quick and trained reflexes. I pulled my gun from my holster and shot his gun clasped hand.

He stood there for a few seconds in awe. In rapid motion, I shot both his arms, shattering his elbows as he dropped the bag.

"Now turn and run, or the next one is though your head," I said grinning.

He ran away as quickly as his feet would stride and his arms just flapped around on his sides. I grabbed the bag and simply walked away. As always though, I took different side streets and alleyways home, zigging and zagging around the blocks to make sure that I wouldn't be caught. It was almost nine when I got home. I turned the bag upside down on my bed, and out fell the wallets in a picture perfect pile. There were probably about sixty or more. I noticed Scott in the corner of my eye and turned to him because I knew he was going to say something.

"What the hell is this all about, Mr. Sandman?"

"What does it look like?" I said sarcastically.

"Did you steal all of those wallets; is that what you've been doing all day?"

"We'll, Scott, yes and no."

"What do you mean, yes and no? Either you stole them or you didn't; tell me the truth!"

"Let me explain to you Scott. It's like this: I happened to have seen a pocket-picker, and after he was done with his picking, I simply picked his pocket. I stole from a thief who stole from honest and hard working people. Simple as that-it's no big deal," I said while putting on some plastic gloves. I started opening the wallets and emptying their cash and only their cash, while Scott looked at me not quite sure knowing what to do or say.

"Don't just stand there, Scott, we have a lot of work to do."

Scott blindly obeyed. I told him to put some gloves on and start emptying the cash. When we were finished, I counted the money: $6,543. I put it in my bedside drawer along with my gun and the ten-thousand already there. That made my net worth, $16,597 plus $300 coming from my government check next week I thought to myself, as Scott caught me chuckling quietly.

"Sandman, I understand what you've done, but what about all of these wallets? Honest living folks usually carry more important things in their wallets besides money."

"I realize that, Scott, and I don't plan on taking anything but the money; everything else stays with their wallets. Now that's where you come in, Scott. I'll buy a bunch of small, folding boxes tomorrow, and you can prepare all of them to be mailed back to their rightful owners. But, I'll insist that you still wear the gloves; how about it Scott, would you do that for me, my friend?"

"If you'll allow me to give their wallets back, then you can count on me," Scott said with resolve and distain.

"Just remember this, Scott; if I hadn't robbed that little pocket-picker, not only would their cash be stolen, but so would their identities, their credit cards, pictures and other important items that can only be cherished by them. The only thing they're losing to me is their money. Think of it as my commission."

Scott put all the wallets into a beer box and sat them next to the kitchen counter to be ready for tomorrow's mail-or so he hoped.

"You know what I was thinking, Scott?"

"Twenty questions; I love this game. Is it made of wood?"

"Yes. What I was going to say is that tomorrow, I wanna buy you a cot."

"Ah . . . Sandman, really, I don't what you to do that. I'll just make a bed on the floor."

"Well, you'll have to stay at least until your leg heals. You can stay here for a few more days, but no longer than a week. Then you'll be on your own again . . . okay?"

"It'll just be for a day or two and then I'll be outta here, Sandman?"

"I want a drink . . . you want a beer and some scotch? I'm going to have some; please join me?"

Scott told me I drank a lot, but he joined me just the same. We ended up sitting at the table saying little or nothing, but just looking out the window into the dark, cold night.

"Would you like me to re-dress your leg?" I asked just to break the strange silence.

"Already did, Sandman, you did a nice job with the stitches."

I turned on the radio and switched the channel to some music. Scott looked as if to be in deep thought. He didn't seem to want to talk much, but occasionally we would exchange a few words, while at most times; again, we just stared out the window. Suddenly, Scott made a comment; "You can see a lot of shit from up here, Sandman."

"Right there is where you used to walk." I said while pointing.

"Yep," he replied.

"Scott, tell me really, where do you get your money?"

"I thought I told you; by begging and pan-handling."

"Scott, with that awesome brain of yours, you could get just about any job you wanted-I don't understand this begging thing you do?"

"Because begging is easy money, Sandman. I don't mind if other people think I'm stupid."

"It might be easy, but surely not enough to live on. Why would you settle for pennies and nickels, when you could have dollar bills instead?"

"Money really hasn't meant that much to me, but knowledge has done for me what money could never do, Sandman. It gives me power . . . and I love that goddamn feeling."

"What power!—Ya asshole . . . I've heard that bullshit before, but for god's sake, Scott; get real! You need money to buy things, you need money to eat, you need money to dress, and you need money for some place to sleep. Money is what you need to survive. It's the shit that makes the world go around. You can't live without it!"

"How do you know I *don't* have money?" he smiled, and then bent his head up to look at me.

"Because you're a fucking bum, Scott. Okay. Now I don't wish to continue this conversation because it's starting to make me angry, and I don't want to be angry, okay?"

"Okay," was Scott's simple reply?

"I think I'm going to call it a night. I want to see Darlene tomorrow and you still need to keep off your feet. Just remember, my poor friend, I'll be waking you tomorrow with another nightmare. Sorry."

Scott didn't say a thing.

I laid on my bed and tried to fall asleep. I opened one eye and saw Scott still drinking and smoking. I watched him for a while and wondered what interesting things were going on inside of his head. Slowly, one eye shut while the other was looking at Scott. Seconds later, I was asleep.

"No, no, no, not there! Oh please, don't drill there, ahhhhhh! Don't do it again, it hurts, not over here, ahhhh, no, no; oh you hurt my ear, please stop. Look at the blood—you've . . . ahhhhh!"

Eleven-thirty in the morning and Mr. Sandman was crawling toward nothing but still trying to get somewhere. I found myself under the table, and then, in the middle of the kitchen floor. It was the brain dream. I took my notebook and scribbled some new notes on a particular page then put it away. The sweating and shaking, the headache and heart pounding were the only senses that convinced me I was still alive. I was alive and coming down from my nightmare horrors to begin another day. I was expecting Scott to be holding me, or shaking me awake, but realized he was nowhere around. He had already left for the day to either study or beg.

I got up and sat at the table. I took some aspirins with the backwash of a warm beer-Scott's beer. I lit a cigarette and stared out the window. I was looking at nothing in particular. I was gazing and bracing myself, while trying to wake up and abandon my dream-mare trance. An hour went by before I started to feel normal. I looked at my watch, which had turned twelve so I decided to get cleaned up. Darlene then came to my mind and that gave me a lot of encouragement. I was looking forward to seeing her today.

After I was fresh and clean, I found a little note from Scott saying he'd be back later. Although I loaned him my spare key, it concerned me that he was already walking on his leg. I let it go-my thoughts were on someone more important to me.

When I left to see Darlene, I took $2,000 with me. But I had a strange feeling that I was forgetting something, or as if I were leaving something precious behind.

On my way to the hotel, I stopped and had breakfast. After eating, I felt much better. A nice breakfast after a night of drinking always made me feel better. I felt human again. Feeling better and more energized, I took a cab to the hotel. After arriving, I went up to the ninth floor and knocked on Darlene's door.

"Who is it?" she inquired through the intercom.

"It's Christopher, Darlene."

"Just a minute, Christopher."

I waited about two minutes when the door slowly opened. There she stood. The swelling in her face was going away, but her eyes were still just a little black and blue. Darlene was slowly turning back into the pretty woman that I first saw on that sidewalk that day.

"Darlene, you're looking much better!" I proclaimed in sincerity.

"You really think my face looks better? I used a little more make-up then I usually do-do you really think it looks okay?"

"You're pretty face is coming back, Darlene, and your bruises are fading away."

She was wearing a very sensuous outfit. She wore a lovely form-fitting, sky-blue pair of slacks with a pinkish-blue fuzzy sweater and low-heeled shoes to match. Her spring jacket was mink white and leather. Because of her gorgeous figure, blonde hair and beautiful deep green eyes, she really stood out from the crowd. Her appearance stood out like a cut diamond in the mix of a pile of crude stones, akin to the masses. I still couldn't believe she was with me.

"Where would you like to go?" I asked.

"You mean, leave the hotel?"

"That's my plan-you need to get out for awhile."

"Christopher, can we first go for a walk in the park? I could use the exercise and a change of scenery-is that okay with you?"

"Whatever you want, you're calling the shots Darlene."

As we stepped outside, she took hold of my hand. I almost swallowed my tongue while at the same time suppressed an erection. I took her hand in a dignified manner and started for the park.

All this time I felt in charge, but yet, initially I found it hard to make chitchat. I thought maybe because her good looks sometimes intimated me-even with the few bruises she had left. But I started to relax as we strolled along the park's paths and smelling the fresh beginning of spring. And I was glad that I had chosen my best outfit to wear today; it made me feel like I was in her class, and not some bum who lived on the wrong side of the tracks. We casually walked in the park as we talked. I felt special and different this day, mainly because I was with Darlene.

"Have you talked to anyone, Darlene?" was my first serious question.

"I called my boss at work," she said politely.

"Oh, right-what did he say?"

"He told me that my job was still waiting. I told him I'd probably be back to work in about a week or so. And again, he completely understood. Thank goodness I haven't lost my job, but more importantly, I'm still alive thanks to you."

"Then I'm happy I helped," I replied.

She looked at me and smiled, and then her expression turned to concern; "My boss told me something else, Christopher."

"What's that?" I asked.

"He told me that my husband called. He asked for my phone number but my boss declined to give it out. He threatened to kill him if he didn't give Vern my phone number where I was staying."

"And then what did your boss say?"

"He told him my phone number to save his life. I told him not to worry about it. Now the bastard has gone as far as to threaten my boss's life. And added to that, Vern knows where I'm staying now."

"I wonder if your boss called the police."

"He said he did, just in case, at least I hope he did."

"Okay. It's time for you to move. Go back to your hotel room and pack your things quickly. I'll be in the lobby paying the bill, and you'll be staying at a different hotel tonight. I'll make the arrangements while you pack. Come on . . . come on; let's get moving," I said as I lightly patted her ass unintentionally and without thinking.

To my surprising delight, she simply turned around, patted my face lightly, and then grabbed my arm as we headed back to the hotel.

About one hour later, we had checked into another nice hotel, and about fifteen minutes after that, we were in Darlene's new suite.

"Not bad," I said and meant it.

We sat down on the edge of the bed. Darlene looked around and then at me.

"Christopher?" she said, "I think you're keeping a secret from me."

My heart started pounding. "And what secret would that be?"

"I think you're rich; aren't you? How else could you afford all this?"

I breathed a silent sigh of relief, "No, Darlene. We've talked about me before-you know who I am. Sometimes I have a little money, and sometimes I don't. But as for now, I have a little money. Which reminds me, Darlene, would you like to go looking for an apartment, or even a house tomorrow?"

"Oh, that would be terrific Christopher!—but I'm afraid I can't afford it now," she said with saddened and glassy eyes.

I suddenly suspected a problem. "No problem, if you can't afford a house, we'll just go look for an apartment. Oh . . . a little snag about something I forgot, Darlene, did you ever call your insurance agent or an attorney? You did call one or the other . . . didn't you?"

"Both, and I even used the lobby phone, but only to discover that my insurance was canceled and the balance refunded, thanks to a little trick that Vern had pulled. I'm not sure how much of the damages I can claim. Even my attorney told me I would be pursing a financial lost cause; in practical terms-chalk it up to life's one, but many little trials," she said in a whining tone, while trying to mimic her asshole attorney.

"He talked in legal jargon and that confused me, but I knew it all came to one simple proclamation: I'm fucked. I had no insurance when all this shit happened, Christopher. I've lost everything. Most of my personal things were lost too. I was so heartbroken that I just cried after I hung up the phone. Because of that fucking bastard, all of my furniture is gone, even my computer. He didn't buy one lousy goddamn thing! Oh! Oh . . . I almost forgot . . . he bought the fucking lava lamp! I don't have any money to replace my things, let alone, the down payment for a new apartment. I've lost, Christopher. I'm a loser. I don't know what's going to happen to me now and I'm scared. For the first time in my life . . . I'm really, really scared, Christopher. What am I going to do?"

After she was through realizing her terrible situation, she began crying steady and hard. I felt sorry for her. I no longer felt intimated by her beauty, but instead, a sudden pain of intense sadness, and a true concern for her well-being. I put my arms around her shoulders and pleaded for her to relax and stop crying.

"Darlene . . . Darlene, don't worry so much; we can work it out. Together we'll find a way. Listen, Darlene . . . listen; I've discovered that to each and every obstacle life throws at me, there's *always* a way to get around them. Trust me; you'll see. You'll get a new place soon-I promise you."

My words calmed her and she stopped crying. I surprised myself. Her shoulders became less round as she straightened her back still sitting on the side of the bed. I rocked her gently with one arm around her shoulder.

"Are you okay now? I'm still sitting beside you-and I haven't run away."

She wiped away a sniffle, "I'll be all right now, thank you; I kind of lost it there for a moment; didn't I?" she giggled uneasily.

"No. Reality kind of shook you up a little, that's all, but I think you'll be okay. Are you feeling better now?"

"Yes, I am, Christopher. Now I feel silly-acting like a damn drama queen. I'm sorry . . . but I'm all right now."

"Good. Let me call room service to order a drink and maybe . . ."

"Let's go out somewhere, Christopher. I need to get out-let's go to a bar or something. I want a few drinks; listen to some occasional music, distraction and talk. But I want to be with you tonight, that's the most important thing. Does that sound stupid to you?"

"No. Not at all, let's go have us a few drinks."

In less then an hour, we found ourselves in the *Broken Ring*; an up-scaled bar filled with many possible victims on a fun, but for some, an unfortunate night. We had a nice time. We got to see a woman slap a man, and then wet him with her drink. I thought to myself that must've been the occasional distraction she was talking about. Strangely, she didn't utter one word about Vern and her dilemma. Instead, she appeared to be focused on returning to her job, "If she still has one," she said. I assured her she would. Other than that, we made small talk as she sat close to me in the booth. Twice, she made some sexual overtones that were hard to dismiss. However, and with sadness mixed with pride, I shied away, trying to be a gentleman and not taking advantage of her over drinks. We spent over five hours in the bar and after she had five drinks, I finally decided to take her home. She was very much obliging and courteous, even though she was slightly drunk.

When our taxi arrived, I walked her to her hotel room. "I'll see you tomorrow," I said, and we kissed each other in the doorway. It was after midnight when I told her good-bye, and then left her standing in the doorway, holding the door open and smiling at me as I looked over my shoulder and gave her a wink.

When I got home, I again put my clothes in their protective plastic covers, and then slipped on some of my everyday rags. I wasn't really surprised to see that Scott wasn't here.

CHAPTER SIX

I went to the refrigerator and found a beer, and then grabbed a fifth from the bathroom. I poured myself a glass of scotch and lit a cigarette. I opened the curtains and looked out that window, half expecting to see Scott walking up and down Steward Street. But he was nowhere to be seen. I took another drink of scotch and another slug of my beer wondering where he'd gone.

A feeling of depression was woven throughout my being. I felt lonely and hungry for conversation again. I already missed Darlene and now I miss Scott. My eyes stared out the window while my mind wandered and worried about my friend now.

I wondered if I would really ever see him again.

I looked by the kitchen counter were the beer box was, but all the wallets filled with their personal items and identifications were gone. I'm betting they'll find their rightful owners.

I finished my beer and went to get a new one when I noticed two little magnets stuck on the refrigerator near the bottom of the door. With the beer and magnets in hand, I returned to the table and played with them. Although I still didn't understand how I thought they might work, I proceeded to try and discover for myself why these crazy little things attract and repel. But after talking with Scott about ions, I was sure it had something to do with atoms, and I even supposed that they were the main principle cause for the behavior of magnets, but I wasn't sure how. I thought that maybe they too, were made of positive and negative ions, but I really wasn't sure how that might be. I put the little magnets down on the table and left them alone. I thought maybe I'd go to the library and look it up, and then decided, fuck it-I'm no genius like Scott.

My attention turned back to the window while I sipped on my beer. The streets were still kind of busy-not with working people, but with the people who live in the night like vampires. My watch read almost two and I thought I might listen to the radio and have a few more drinks. I was watching out the window when suddenly, two cars collided in the intersection. It was just another of the many fender-benders I've seen before.

One man got out of his car and started talking to the other who then also got out. They stood there like gentleman exchanging phone numbers and other information. Within a few minutes a police car stopped with his lights flashing. Some of the people who were in the crowd watching walked away swiftly when the cops arrived. One car was towed away and the other was able to drive off okay. In the span of an hour, everything was back to normal, whatever normal was.

I continued to drink and smoke and play with the radio dial. I came across the national news and listened to the politics, and then the weather and half-hour news program about a new telescope that NASA was building which would see to the edge of the universe. The edge of the universe for God's sake! I wondered if NASA might see a big pair of eyes staring right back at them. I had to smile to myself at such a ridiculous thought. When the space news was over, I sat back with my legs stretched out and my arms crossed. The stuff that they talked about was truly amazing. It made me think of Scott; he would have enjoyed the program. I was thinking of sleep, but for some terrible reason, I didn't sleep that night. I continued drinking until my dark side came alive. I was buzzed and feeling very fiendish. The bad thoughts brought out the worst in me and I wondered how evil I really was inside.

Was I capable of living in two worlds: One of hate and the other of love?

Could I somehow change for the better, and in the near future, or was I only as good as all those other shits who walk down there at night doing wrongful deeds? My thoughts answered my questions.

I drank some more until I thought of an old scheme; something I've always wanted to do but never did, but now I was going to do it. It was time to take another walk-another walk into the night.

It wasn't very cold outside, so I decided to wear one of my new coats; besides, they were the only ones that had room for my shoulder holster and gun. I decided to take a walk down Steward Street going east. The far east side of town is considered the worst part of the city because of the violence and gang activity. I went out looking for trouble and if there was money involved, that would be all the better. I was about ten blocks east of my apartment when I came across a gang of street thugs. They didn't frighten me in the least because I had my super buzz on, and I was hoping they would approach me. It was just a matter of a minute or two before my wish came true.

"Say, dude. What's up?" one of the punks said.

"Just taking a little walk; enjoying the night air," I replied slowly.

"Hey, man, you got a cigarette?" he said.

"Sorry, fellows, but I don't smoke," I said, exhaling from a cigarette.

"Hey Spike? Get this. The motherfucker thinks he's being funny."

"Cool, dude." Spike said.

"But I still want a cigarette, and while we're at it, I think we'll kick your fucking ass too." He spoke again.

I took a drag on my cigarette and looked at the guy who was talking. All included, there were six punks in this little gang. My back was still open and they faced me from about five feet away. I decided to make my move and started talking.

"I'm going to tell you straight up, all of you little nim-rods, you're fucking with the wrong man tonight. I'm going to jack you up and smash you down because I'm in a killing mood now. *You* want to kick my ass? Swell, but I'm going to see your blood on my hands when it's all through. You boys don't understand; you're messing with a crazed and desperate man. I'm the fucking Sandman!" I yelled my words and laughed simultaneously like a possessed demon.

They stood still and captive while I was yelling and laughing, and within that very instant, I quickly drew my gun and shot the punk who was talking in the face. Not to kill him, but just to make him ugly for the rest of his life. I took turns pointing my gun in the general directions of the other five.

"There goes one, any other of you little soldiers have any guns?"

"No mister! Please . . . just let us go?" one of them whined.

"No guns? Then I guess all we have are knives tonight. Honestly, how do you pukes make any money? Okay. Now I want all your cash because today is my payday."

"Please, don't go crazy with the gun mister," the skinny one said.

"Please, sir, don't kill us," the biggest one begged.

"Shut up!" I screamed. I felt so angry that it was making my face sweat. I've never noticed it before, but I suddenly became aware that my teeth were gritting down hard, and I was smiling. We were standing on the sidewalk and I gestured with my gun for them to step between two buildings. The one I shot was still lying on the ground holding his face and crying. I knew he would live because I just shot him in the mouth. If he manages to survive on the streets, well, he'll just grow older with fewer teeth, and less jawbone.

"All of you, move over there and grab your fucking friend," I demanded.

They all did what I said, and now we were hidden from the streets when I decided it was going to be a payday for me.

"I want all you little cock sucks to give your pal on the end all your money, and don't forget your bloody friend."

They reached into their pockets and pulled out their money and gave it to the kid who was standing at the end and closest to me.

"Now, you, give me the money, slowly; and if I see one small move, I swear to God I'll kill you all!" It started raining cold droplets of water.

The punk who had all their money slowly reached over and handed it to me. I could tell from the look in his eyes that this was the worst situation that they'd ever been in-an encouraging sign for me.

"Okay, fellows, that was very nice of you. Now I want all of you to take your clothes off-even your shoes and socks. Throw them in a pile over there."

Within less than a minute, they stood in front of me, five young men, naked, wet and frightened like the little boys they were.

I just had to have a couple of moments of fun with these twits before I was done.

"You know, you kids don't fool me; never did; never will. You all want to be the toughest *man* on the street, the biggest and fearsome gang in the neighborhood. But you're all losers-and you want to know why? Because there'll always be people like me. Now, you all want to be big men-right? Well . . . big men have big dicks. I want all of you to look at each other and tell me who has the biggest dick."

They all looked at me in astonishment. They were confused and puzzled while I was laughing, and shouting, and then belted out as loud as I could, "Come on you little fuckers, who has the biggest dick?"

"Spike does," they all said pointing. I had to laugh.

"One last command, you cute little bitches. Turn sideways, bend down and spread those little ass cheeks. That's good . . . very good. You guys must've played this little game before. Now, I want you all to kiss your buddies assholes-and I'm not talking about butt cheeks. Do it! And I mean now goddamn it or I'll blow all you sons of bitches away! I want to see some brown shit circles around those lips, okay. Now, do an about face and kiss your other friends ass in the same way. Come on girls! I know you like it. I want to see some fucking tongue action going on! Give your buddies a good old wet one, you stupid mother fuckers. That was a very lovely display, fagots-I wished I had a camera. Oh, by the way, I know the ground's a little muddy but I didn't plan that. Now, lay down on your stomachs, and don't look up, or I swear to God I'll shoot your eyes out!" I yelled and spit like a boot-camp drill sergeant.

Again, they did what I said. I shot a round in the skinny ones ass to keep them scared. After he fell to the ground I turned around and swiftly ran away. I went between different streets, zigzagging as always until I was far enough away to slow down. It took me about half an hour to make it back to my apartment.

I closed the door and fell on my bed laughing like some kind of college frat. I couldn't help it. I always wanted to do something like that to a gang of street punks. I began to quit laughing and started counting the money: $463. It really doesn't pay enough to rob these little bastards, especially when I know there's a slight chance the little fuckers could win, and then I'd be dead. That would be the shits. But I also wanted to do it for spite. I walked into the bathroom and took a shower while thinking of Scott and his leg; I hoped that it was them who cut Scott's leg that I fucked over tonight. As I washed off the musty odor of soiled and rain soak clothes, I thought with pleasure, this one's for you Scott, this one's especially for you."

After my shower, I put on a bathrobe. It was just another rag I occasionally wear. It's about five years old and has a few burn-spots from my cigarette ashes. I sat at the table and took a gulp of my scotch. It was still dark outside and I wondered what time it was. I looked at my watch-five in the morning.

I grabbed a beer from the refrigerator and drank some more thinking what a stupid stunt I had just performed. Something I always wanted to do, but never again. I took the money from my bedside drawer and added to whatever I had left. I counted about $14,300. Ghetto income.

Sometimes you're up; most times you're down. I drank one more beer and about three cigarettes. I felt a twinge of guilt in my gut and it just didn't sit right. I decided that I'd have to tone down my drinking and nightly walks.

I'm going too fast. Besides, if I want Darlene to like me, than the sooner I must realize that I have to change. Become an honest working drone of some sort. Assuming that is, if we stay together.

I was becoming tired, and was trying to put it off. But the inevitable can't be put off forever-it was going on six in the morning and time to lie down so Mr. Sandman can get some sleep.

I immediately grabbed the edge of the bed and hung on for dear life. I had a terrible but sensational feeling of falling. But the most horrifying things about my dreams are; they feel so goddamn real. Most of the time, I'm actually feeling as if I'm going to die, and dramatically ugly. Only a few people have seen that once in their lifetime, and only once, but not me. I opened my eyes quickly and began blinking the sweat out of them. I quickly came out of the nightmare with a heaving chest.

Mr. Sandman was awake. I looked at my watch, which read twelve noon and six hours of sleep. I took my aspirins and grabbed a beer and looked at the scotch that was still on the table. I lit a cigarette, shot a small swallow of scotch down, and chased it with beer. The sweat was dripping off my forehead and on the table. I opened the window to let some breeze come in. It was cool and dry, and it felt good. Just on the off chance, I looked out the window to see if I could find Scott, but Scott was nowhere to be seen.

This was the first time I had opened that window in about eight months. I was starting to hear and smell the sounds of the city streets again. I heard the faint sound of cars and a few honking horns. I heard people talking, laughing, and at a distance, some arguing somewhere below. I smelled food cooking from a window. But I could also smell that odor again; air contaminated with fog and soot from the exhaust of combustible engines and factories. At this time of the year, the noxious vapors take longer to disappear.

I knew it was going to be a long day today. I got up and took $4,500 from my drawer. My remaining total was $9,800. It seems like a lot, but in the city, it's not. God how I wished I were a normal person; like someone who had a steady job and a nice place to call home. Someone who could love me and stand beside me, and someone I could love in return. I had to stop thinking of these thoughts because it hurt too much and besides, it was starting to remind me of Darlene.

If she knew the truth about me, she'd never want to see me again. In reality, I was nothing more than just another goddamned fucking bum and a thief. I literally started to cry, but strangely near the end, I found myself laughing with the thought of the pocket-picker.

After a cold shower and a few more aspirins, I put a nice outfit on to be seen with Darlene. Today, I wanted to look for a new place for her to live.

I left with $4,500 in my pocket and a lot of dumb pride in my soul. I bought a newspaper, and then flagged down a taxi to Darlene's hotel.

"Is that you, Christopher?" her voice resonated through the intercom.

"It's me, Darlene."

When she opened the door, she looked even better then yesterday. Most all her injuries were gone, and with practically no noticeable signs that she'd been beaten. She was wearing some casual jeans and a tan-brown sweater; this was the first time I saw her hair in a ponytail.

"Are you ready?" I asked.

"Ready for what?" she smiled.

"To look for a new apartment!"

"Christopher, I told you yesterday, I don't have the money . . ."

"Don't concern yourself-I'll take care of the deposit and the first months' rent. Why don't you checkout some ads in today's paper, and then we'll go see how they look?" I handed her the newspaper.

"Are you serious, Christopher?" she replied wide-eyed with a grin.

"As serious as a deadly snake bite . . . does that answer your question?" I said trying to add some humor.

Darlene flipped to the real estate rentals pages and scoured them vigorously. She circled a few ads with a red pen. "Can we look at these apartments today?" she asked in a girl-child voice and expression.

"Whatever you want, but before we go looking, can we get a bite to eat first? I'm hungry."

"That would be nice, I haven't eaten anything today either," she said.

We took a cab to a west side restaurant where she had eaten once before: *The Bellwort Exclusive*. It was a very fashionable restaurant that catered mostly to the rich. Although I had a little trouble with the menu and the wine selection, everything else went just fine. We talked as we nibbled on bread sticks, crackers and cheese-and when our cuisine arrived, we conversed between bites.

"I talked to my attorney this morning. He's going to draw up some divorce papers so I can get rid of Vern."

"That's a step in the right direction," I concurred.

"I wonder what he's doing?" she said in a strange way expressing more curiosity than concern.

"I'll tell you," I spoke up, "he's living on the streets."

"Think so?" she asked.

"Let me answer that question with another question. Do you think he's living in some luxurious hotel, or a new fancy apartment, or even in an expensive home?"

"No." she simply answered.

"That's right," I said, and then added, "My guess would be, that if he's found an abandoned building to live in, at best, that's about as good as it'll get for him. How's your food?"

"Very good-I love this salmon steak-yum, do you want a taste?"

"No thanks. Have you decided which place you'd like to look at first?"

The apartments where she wanted to live were on the west side of town. We took a cab to our first appointment and inspected the apartment. Darlene didn't care much for that particular one. And when we arrived at the second, Darlene immediately fell in love with the fancy brick exterior. We inspected the inside with the agent who talked non-stop. She loved the fireplace, kitchen, and the fact that it had her own private bathroom off from her bedroom and a half bath from the hall for guests.

I asked the real estate agent what the rent was and she told me, $1,250 a month plus utilities. When I asked her what the deposit was, she replied twice the monthly rent.

"Can you afford over $1,200 a month, Darlene?"

"I believe so. The other apartment was about $960 a month plus utilities and it was a dump."

I excused ourselves from the agent and we walked over to a quite corner to talk. "Darlene?" I asked, "How much do you make?"

"I take home $2,350 every two weeks."

"Well then, it's up to you. Did you want to take a look at the third place?"

"No, Christopher, I want this place."

"Very well," I said. "I'll loan you the down payment and you can sign the lease. That'd save us a step so you can move in tomorrow."

"I swear to God, Christopher, I'll repay you as soon as I can get a loan."

"Okay. I believe you, Darlene; but enough said for now."

Her credit was outstanding and that kind of surprised me. How she managed to keep a clean credit while living with her husband had to take some clever imagination and smart thinking. With her good credit, she got the apartment.

Tomorrow, she wanted to see if there was anything in the old apartment that was salvageable. I kind of had my doubts about that idea, but soon decided that Vern would be nowhere around anyhow. Then she wanted to go to some furniture stores and buy some new furniture for her apartment. I was becoming strongly aware of that old concept: give an inch and they'll take a mile. Was she starting to take advantage of me? When we arrived back at the hotel, I told her that I'd see her tomorrow. It was a little after seven when I got back to my dump. Something felt wrong. I wasn't sure if I was feeling glad, sad or mad.

I took off my new clothes and put on some casual shit and sat at the table with a cold beer and scotch. My mind felt null and my life seemed worthless. Here I was, dolling out cash today like I was some rich motherfucker who owned a fleet of jets. I'm just a lousy, damned, good for nothing, fucking bum! Who in the hell am I trying to kid? My damn rent is $250 a month including utilities. I took a big slug of scotch and downed my beer. I felt like getting blasted. I began to wonder more intensely if Darlene was taking me for a fool. She must know that I'm not made out of money? She knows I don't have a steady career job. Why is it not in her mind, *my* financial concerns and condition? My mind would be busy tonight.

My thought's kept turning over the same old issues. I kept trying to put myself in her place. What would I be thinking? How would I be reacting? Was she no longer fearful that Vern would find her? Did she find herself safe with me? Did she have any real feelings for me? Did she kindle a little spark in her heart as I do for her? Maybe I gave her the impression that I was too rich. But that couldn't be; she saw where I lived. I wished Scott were here to talk with. He could offer up some good advice I'm sure-I felt so alone.

All these thoughts were going through my mind, over and over and over. What am I going to do when the money runs out-take more walks in the night?

Maybe I could find a job somewhere, but where? I can't sleep regular hours. I'd have to stop drinking and slow down on my smoking. I may even have to move. Maybe I can pick up and move somewhere far away. I was starting to develop a headache with all these issues. My mind craved serenity.

I got up and grabbed a cold beer and a new fifth of scotch from the bathroom. I poured a drink and gulped down three quick swallows, and then I chased that with half a beer. I lit a cigarette and looked out my window. I always expected to see Scott, but I never did anymore. I wondered where he went.

It was a little after eleven. I was on my sixth beer and the bottle was just about empty. There was one last bottle of scotch in the bathroom and a little over a six pack of beer in the refrigerator. I would have to find some time tomorrow to make a stop at Billy's Store and stock up again. After a couple hours, I was feeling better because I could feel that good old, *fuck-it* attitude inside. Like many say; drinking is only a temporary solution, but I felt to hell with it, it was the only solution that was going to temporarily solve my problems tonight. Come to think about it, every night. But what the hell, my life is damned anyway. I might as well sprout wings and fly to hell. I sat at the table and had a little pity-party for myself.

Intertwined with thinking, and listening to the radio, I was also watching out the window-for hours and in a drunken, stupid stupor. Finally, I made the courageous attempt to go to bed.

I couldn't remember what time I went to bed last night, but Mr. Sandman was now awake with the sweats, headache and other shit. But something different happened for a change. For the very first time ever, I could not remember the nightmare I was having. Usually I can't get it out of my mind until after an hour or two, but this time, I had no idea what horror awoke me? To say the least, I was surprised, even kind of glad. Maybe this curse is starting to go away. It was eleven-thirty in the morning. I took my time making myself some eggs with toast and coffee. After I went into the bathroom for my morning clean up, I suddenly couldn't remember what I did with my cash from yesterday.

I breathed a sigh of relief when I discovered the money in a pocket of my new pair of pants that I had worn yesterday; I found about $800.

I put on a new outfit, grabbed the money, my holster and gun, and then left for the hotel.

When I arrived at her door, I didn't knock but instead, just kind of stood there for a few minutes feeling like I was just an illusion in her life. I knew deep in my heart that I wasn't good enough for her. Eventually my facade would peel away and expose me for what I really am-something lower than a snail's ass. Finally, I held my head up, squared my shoulders and took a deep breath and knocked on her door.

"Is that you Christopher?" she said near the door.

"Yes Darlene, it's me."

"Just a minute," I heard her say from a distance this time.

After a few minutes, the door opened and there she stood, like a princess from a majestic world. She was wearing her hair in a fancy braid and dressed in white form fitting jeans, with a golden-laced top, white socks and tennis shoes tightened with gold laces.

As always, I told her how nice she looked. I then asked her if she would like to get something to eat first, but she was too eager for that. She wanted to go to the furniture stores and pick out some new things for her apartment. We took a taxi to a fancy furniture store on the north side of town.

She told me she really had no other option but to purchase her new furniture on her credit cards. I turned around and breathed a big sigh of relief. She bought all kinds of things. Things that make a home feel like it should be. All kinds of things that I've never known—as I've never lived in a real home. She had spent over $15,000 for new furniture which was to be delivered tomorrow sometime in the afternoon. That worked out good because tonight was her last night's stay in the hotel.

After we left the store, we got something to eat. We ate at a regular family restaurant this time. She told me that she was very excited about getting her life back together. She wouldn't have to support her husband anymore. She could save more money. Bills would be easier to pay. She would be able to buy nicer clothes and other things. She talked on and on and I couldn't help but to smile when I saw her so happy. All the while my heart lay heavy, but my soul, if I should have one, took comfort in her joy.

It was about six o'clock when we arrived back at the hotel. She invited me in for a drink and to sit and relax for a while. I did and we talked about tomorrow and how she couldn't wait to get out of the hotel room and into her new place. I didn't want her to know that I was carrying a gun, so I left my sport coat on.

We were sitting on a sofa and the talking gradually slowed to silence. Without warning, she leaned over and kissed me on my lips and I kissed her back.

"What was that for?" I asked her afterward and feeling momentary exhilaration.

"Because I like you very much."

"That was a very nice way of showing it. I like you too, Darlene."

"Christopher, you're a very nice man but please don't get upset if I tell you I'm starting to like you more and more.

You're starting to feel special to me, more than just a friend."

"Darlene, in the beginning I was not helping you for love, but I must say now, I like you very much too-my feelings for you have grown."

That exchange of words is when our lives pursued a new change in our relationship. We were boyfriend and girlfriend now, and I was thrilled. We were flirtatious with each other as we talked, and some of our conversations were private and intimate. But we also talked about other things. We talked a little while about relationships, and friends. Things we liked and disliked. We were starting to get to know each other more then just friends and she also showed me a whole new quality in herself that I liked.

"Darlene . . . as much as I'm enjoying myself, I really should be going. I then told her a little innocent fib, "I promised to give a friend some help . . ."

"Is your friend a man, or a woman?"

"Does it matter?" I asked smiling.

She gave me a little *love-push* and said, "I'll see you tomorrow," and then gave me a kiss and added, "I love you, Christopher."

"God, what a relief . . . I love you too. Tell you what. I'll visit you at your new apartment later on in the afternoon-are you going to be around?"

"If you're coming I will. I'd love to see you tomorrow, Christopher."

"It's a date then. Oh. Do you have enough money for a taxi to get home in the morning?"

"Yes. Thank you so much, Christopher, and don't forget to come see me."

"I won't, Darlene."

I took a cab to Billy's Store to re-supply my habit. I wanted to pick up another fifth or two and a twelve pack of beer.

"Hey! There he is, Mr. Sandman. How you doing with yourself?" Billy said while patting my back.

"I'm doing pretty good, Billy. I just stopped in to buy some scotch and beer. How's business been doing?"

"Business is slowing a little, but I'm not worried, Sandman. One day I get some customers, and other days I don't get as much. But it's still paying my bills. Are you still having those terrible dreams?"

"Yeah, but something strange happened to me when I woke up today. I knew that a nightmare woke me up, but when I opened my eyes, I couldn't remember the dream. That's never happened before, Billy."

"Well, maybe God is healing your head; you know I pray for you, Sandman. Maybe He's listening to my prayers?"

"Yeah, maybe he is Billy," I said paying little attention, "Do you have a couple fifths of my scotch?"

"I think so, let me see," Billy said as he bent down searching for my particular brand of scotch. I went to the cooler and got a twelve pack and sat it on top of the counter. "I see only one here, Sandman. I have more in the back room; could you go and get it for me? It should be on the shelf to your right."

"No problem, Billy."

I walked back into the storage room and walked to the shelves to the right. I picked up a bottle when I heard some customers come in. I turned around and started for the door when I heard the words that would strike terror and fear into anyone's heart.

CHAPTER SEVEN

"This is a hold up! Give me all your money or I'll blow your goddamn fucking head off!"

I approached the door while setting the bottle of scotch down behind me quietly. From the reflections of the coolers and freezers, I saw two men holding guns. In a quick but soundless motion, I bent down to my knees and spied them from the bottom corner door jam near the floor. This little trick makes me stealthier and less likely to be seen. Suddenly, I heard a gun shot striking something metal.

"Open your cash register *and* your safe or you're a dead man you old son of a bitch!"

"Oh my God! Okay, okay, here's the money from my cash register. I'll open the safe but it's below the register; I have to bend down to open it-please don't shoot me," I heard Billy say. I felt sorry for him because of the terror and the frightfulness in his voice.

"You just go ahead and do that; but nothing stupid or I'll waste your old fat ass right here and now!"

I peaked around near the bottom of the door frame and removed my gun. Billy never knew that I carried one. I quickly stood up and stepped into the doorway and fired two shots. I pulled the trigger twice, swiftly shooting one man, then the other. Both their heads went backwards while they were still standing. Blood splattered behind them and then they dropped like marionette puppets whose strings had just been cut. I shot them both, dead center in the throat.

I ran up to them and saw that their heads were basically severed-instant kill, or so I thought. As I looked at one of the men whose head was cocked slightly sideways, I noticed that his blank eyes suddenly came alive and looked up at me. I saw his lips moving, but there was no sound-he had no voice box. In his mind, I'm sure he thought he was talking, but nothing uttered from his lips.

I saw his chest still going up and down, and I could hear the blood and air bubbles coming out of his raged neck. He did this for about fifteen seconds until his eyes quite moving, and his lips stopped moving as well as his chest. The last thing I saw were those blood bubbles coming out of his neck as he exhaled his last dying breath. I've never seen death that horrific before, but I didn't have much time to think about now. I turned around to look for Billy to see if he was okay. Thank God that through all the commotion, no customers walked in.

"Holy shit, Sandman, what happened? How did they die? Did somebody shoot them? I heard nothing . . . all I saw was their necks exploding and they just dropped dead!" Billy blabbed quickly while waving his hands aimlessly.

"I shot them Billy," I said without concern. "Go lock your front door, and be careful not to step in any blood. I have a plan so we have to move fast-now listen carefully. I'm going to step in some blood and run out the front door; make the police think that the tracks were made by some unknown killer."

Billy was locking the front door while I kept talking. I didn't want any customers coming in and seeing this mess. As Billy passed near me I smelled something awful-he had shit his pants-I ignored the smell.

"When I leave, you immediately call 911," I said.

"What do I tell the police, Sandman?" he answered concernedly.

"Tell the police that the two guys' who were robbing you, went into a panic when suddenly a third man walked in. Tell the police that they all seemed to know each other and the third guy who walked in was also in on the deal. But abruptly, they started to argue when the third guy stepped in front and shot them both, just as you see them know."

"But there was no third guy, Sandman."

"The police don't know that. Tell them that the third guy was angry and yelling at the other two-which you didn't understand. Billy, listen to me-these two guys were going to kill you! Trust me, I know what I'm saying. I know how they think."

"Oh fuck, Sandman, this is such a terrible thing, I'm so nervous I don't know what to do?"

"Billy, just do as I said . . . please! I just killed two guys for Christ sake; do you know what that makes me? I have to get out of this store-now! Tell the police what I told you; they won't question your story. Now let's get this shit over with. Just stick with the story I told you. They'll have nothing else to work with, except my bloody footprints. But I'll take care of that; now do as I said."

"All right, Sandman, I'll do it." He replied extremely shook up, and which might be to my advantage.

"Okay, Billy, unlock the door now because I'm getting the hell out of here. Remember, as soon as I leave, shut the door and call 911."

I ran down the sidewalk, and then cut across a field while stopping only momentary to pull off my shoes and socks. Then I ran non-stop through different blocks, sometimes back tracking and making confusing trails and circles. When I saw the opportunity, I scuffled my feet in water, oil, grease and even garbage beside the streets. I jumped fences and ran through alleyways then simply started to walk in random directions.

When I was a few miles from my apartment, I happened upon a few bums standing beside an old oil drum with a fire for heat. As I walked by, I threw my socks and shoes in with little rejection from the derelicts standing by their fire drum. As a matter of fact, I heard one of them say, "Thank you," as I kept on walking away.

When I was out of their sight, it was my belief, that if anyone spotted me, they'd think that I was just another bum on the street. I eventually made my way back home without incident.

When I arrived, the very first thing I did was scrub my feet with water, soap and bleach. Then I opened a bottle of scotch and belted down about five swallows. My nerves felt like they were stretch around and on the outside of my skin. I was shaking and frightened to the point of shear panic. Any little sound I heard, I found myself recoiling in action. I got up and grabbed a beer and noticed that I had three cigarettes smoking at the same time. My mind was in complete chaos. Momentarily, I couldn't focus or think-my brain was full of irrational thoughts. But I did understand one thing-I was now a killer-a murderer.

I kept telling myself that I shot those two men in Billy's store for his defense. Have I gone too far this time, I wondered? But I knew in my gut, that they would have killed Billy and I just couldn't take that chance. I would have felt like a coward and wouldn't be able to look at myself in the mirror. I hoped it was the right thing to do, and from my perspective, it was my only choice. It had to be the right thing. I just happened to be in the right place at the right time or Billy would be dead. I was drinking like there would be no tomorrow for me. I was getting drunk and glad of it. I wondered what Billy told the police. I looked at myself wearing a nice outfit that was now torn, ripped and stained with blood, and of course, barefooted. I took my bloody clothes off in the shower stall and threw them in a plastic bag, and then took a nice warm shower.

Afterwards, I disposed of my tattered outfit by opening a manhole cover and dropping them down the city sewer. I didn't want to be caught by the police by leaving an obvious trail.

Although my apartment wasn't much bigger than a prison cell, I still had the freedom to move about and come and go as I pleased. If the police ever caught me and threw me in a prison cell, I knew I'd go completely insane-and I mean completely. I fell asleep drunk with my head on the table lying in a puddle of beer.

When I woke up, I found myself on the bathroom floor yelling and screaming from my nightmares. I hoped the blanket wrapped around my mouth muffled them, which I somehow managed to come across last night in my sleep. I had a horrible hang over and felt for sure like I was going to die. Added to my alcoholic condition, I also woke up with the sweats and shakes, a bad headache and the memories of last night. I managed to lift myself up and stagger to the table. Now, even my living life seemed like a nightmare. For just a moment, I thought what happened last night was just another horrendous dream, but my waking senses told me otherwise. It was the most dreadful feeling-a dream like state of the mind. Something that only a killer could describe. My watch read three-fifteen in the afternoon. I quickly put on some shoes, grabbed some change and went downstairs to buy a newspaper from the corner stand. Tucking the paper under my arm, I hurried upstairs to see if what I'd done was in the *Daily News*. I eagerly unfolded the paper and laid it flat in front of me.

"Holy shit. The damn thing made it to the front page!" I shrieked out loud. There was a colored pictured of the front of Billy's store and another picture of the two dead men with rags over their heads and shoulders. It was a typical picture that newspapers have shown before. But this time, I knew the dead bodies were those of evil men. Their arms and legs were sticking out from under rags with pools of blood baptizing their sinful souls. I began reading the article.

> "There was a bizarre shooting at *Billy's Party Store* on the east side of the city along Steward Street last night when the owner called 911. When police arrived, they found two men shot to death while trying to rob the store. It was an exceptionally gruesome scene, and police are still searching for more details. Police are investigating this strange incident when the store owner, William (Billy) Richers told police that two men walked in his store and demanded money while a third man stood outside. One of the robbers fired his gun, which struck the cash register. The third man, who was believed to be involved, came in and apparently started to argue with the other two.
>
> According to Mr. Richers, the third man is suspected to have shot his companions before running out of the store. Other than the storeowner, there appears to be no other wittiness to this bizarre crime. However, if anyone has any information, they are asked to call the police. Mr. Richers was taken to the hospital with apparent chest pains. The two dead men were identified as Vern Osmon and Chuck Henderson. Please see *shootings* on page B27 for more details."

I turned to page B27 and continued to read, but there was really nothing more said about the police investigation except that the shooting was a robbery that probably had gone bad. It also said that Billy could not give an adequate description of the third suspect and that the case could take a while to solve. After I read the article, I breathed a sigh of relief. I thought that maybe this time, justice would be served without any outstanding circumstances. I also noticed that one of the men I had shot was Darlene's husband. I had no idea at the time that it was he who had mouthed his unspoken words.

By four in the afternoon, I was calm enough to go into the bathroom and clean up. I wanted to keep my date with Darlene because I was also curious to see what her reaction would be. After my shower, I put on a set of new clothes and took a cab to Darlene's new apartment. I arrived a little after five and just before I knocked on her door, she swung it open.

"How'd you know it was me?" I said.

"I saw you get out of the cab; besides, no one else knows where I live, especially my ass of a husband," she said distraughtly.

At that moment, it was obvious she knew of her husband's death. I put forth my best effort to pretend that everything was okay-that is, until I got a better look at her face. "Did something happen to Vern?" I ask pretentiously.

"Yes. Now I have the added financial burden of burying that stupid son of a bitch," she grumbled angrily.

"You mean he's dead?" I asked, again pretending an inquisitive face.

"Vern was shot to death last night trying to rob a store."

"Really?-are you okay?"

"Hell yes, I couldn't be better. I knew it would be just a matter of time before something like this was going to happen."

"Gosh damn, Darlene, I'm not really sure what to say."

"Don't say anything-he got what he deserved-the stupid dumb bastard."

"May I come in?"

"Oh, I'm sorry, Christopher; yes, please come in."

"Hey, this place is looking very nice-I like it."

"Thank you, but I wished I could be as happy about it as I was earlier this morning. I went down to the police station after receiving the news."

"What did they say?" I asked with more hidden interest.

"I had to identify the body and it was so awful, Christopher. They told me that Vern was shot and killed during an attempted robbery last night-a robbery they believed had gone bad. The police told me that the investigating team said it would probably be an open and shut case due to the fact that they were previously sought after criminals with many warrants.

Their deed was their own doing they said, and also added that, many of these type of killings go unsolved.

Everything seemed to confirm that the storeowner was telling the truth. The owner claimed that a third person walked in and an argument ensued. The police are speculating that the third person might have been a lookout man. They believe it was he who shot Vern and his buddy, but without further evidence, they have no idea who it might be.

Vern had a rap sheet three feet long and they were going to see if they could link him with other felons he might have known. They had me sign a statement, which in essence, said: I had no involvement and was not aware of the incident before or at the time the crime took place. After I signed the paper, I was free to pick up Vern's body from the city morgue.

No autopsies were done because it was obvious what killed him. The physician at the morgue only drew blood. Apparently, Vern and his dead buddy, Chuck, had a considerable amount of crack and alcohol in their systems. But, you know what really burns my ass, Christopher?" she said, and before I was able to ask her, she was already giving me the answer.

"What really pisses me off is that creep busted up my things, and then sticks me with the expense of having to bury him. Now I'm stuck with that added expense-the motherfucker. I'm happy . . . no, I'm ecstatic the asshole's dead!" she said, and started crying. "I can't afford all this now. It'll cost me thousands. I don't have that kind of money, Christopher, and we didn't carry any life insurance, even though it wouldn't have helped."

"Darlene. I have an idea?"

"What?"

"You don't have to bury him. Donate his body to science."

"Can I do that?" she asked, wiping away her tears.

"I think you most certainly can. Of course, I don't believe you'd receive any money, but universities are always in need of human cadavers for their anatomy and medical classes. Besides, Vern will actually be doing something good for a change-wouldn't you agree?"

"Well . . . yes, but what'll I have to do?"

"Well . . . I'm not exactly sure, but you could start by talking to the police. They could probably tell you what kind of legal permissions and actions you'd need to take and maybe some other assorted things. If there are no complications, I suppose the next step would be to call the university. If they're interested, and I'm sure they would be, have them call the police. I would imagine that they would take care of the rest; you know, picking the body up from the morgue and so on."

"Do you think it would be okay if I called them today?"

"I wouldn't worry about it today, the police may still want the body for something and besides, it's getting too late-I'd give it another day. You may have to do some running around and sign some papers, but my guess is that it shouldn't take more than half a day. Then, after all this shit is over, you can get on and back to your life; your new apartment and return to work."

"I'm not looking forward to tomorrow, Christopher, but you have a way of making things seem much easier. I'm so glad I helped you that day you spilled your groceries. The only other *friends* I have are at the office, but they couldn't help me the way you have. I'm so fortunate that I met you."

"I feel the same, Darlene. Before I met you, my life seemed so unfair, but now, it seems so much more fair, and a whole lot warmer too.

"That's so sweet of you to say."

"And I mean every word of it. Hey, this place is looking nice . . . would you give me a little tour? I'd like to see it with your new furniture?—And afterward, if you'd like, we'll get a bite to eat."

Darlene gave me a smile and started showing me the rooms and some of the things she had managed to display, that is, the one's that Vern didn't break. The furniture store had already delivered everything and apparently had arranged things by Darlene's instructions. Her new TV was on, but the sound was muted. She had a lovely home-a new home she could call her own.

When she showed me her bedroom, it resembled a woman's bedroom exactly like the pictures I've seen in some home magazines. Her bedroom looked girlish and nothing like a man's bedroom-especially mine. I noticed she had made the bed and set out some girlie stuff, like stuffed animals, panties and bras, jewelry boxes and two pretty vases with flowers. I spotted a little ballerina music platform with a round base, and when I wound it up and sat it down; the music was as pretty as Darlene. She also had some cosmetics and other assorted things. She had managed to hang a few pictures, and I was kind of surprised that I actually liked them.

I've never been in a woman's bedroom before; there was something very elegant, and very special and beautiful in her room. By comparison, it made me feel dirty and sloppy, like an old, scabby, wet dog that shouldn't be in here. But the strange thing I remembered, and even though the bedroom was new, I found that if I took a deep breath, I could smell the very essence of her attractive hormones and womanhood, and I cherished every second.

"Christopher, come in and look at my kitchen," she said, and I followed her out of the bedroom while sneaking one last peek and smell.

"Wow. This is nice! You even have your new dining table settings displayed."

"And look at all the cupboard space I have!" she exclaimed excitedly, only as another woman could understand. I've heard that women just love lots of cupboard space.

She bounced around the kitchen showing me all kinds of things. She showed me her new stove, oven, refrigerator, microwave, sink, garbage disposal and her kitchen island. She even managed to set out some of her kitchen displays. She also showed me the family room, with the impressive gas fireplace and new chairs. The last was the half bath off from the hallway. She indeed had a nice home. The carpeting varied in texture and color from room to room, and it made the apartment feel even more appealing and spacious, changing and not staying the same.

"Darlene; I'm impressed, now this is what I'd call a home. I'm happy for you. But how did you get your personal things?"

"When the movers finished moving all my furniture, I told them I'd buy'em each a six pack if they would take me to the old apartment and get some of my things. They agreed, and so I was able to get the rest of my unbroken things. When I got back, I set some things out for display; how do you think I did?" she asked me and smiled.

"That was pretty clever," I said.

"I thought so too," she replied. "And after a few hours, I took a break and turned on the television. The local news was on and that's when I heard about Vern. I quit unpacking and took a cab to the police to tell them he was my husband."

"Are you sure you're okay, Darlene?"

"I'm sure; I'll be fine. I didn't love Vern anymore and now I don't have to worry about answering to him, or get beaten by him. From the day he started to hang out with those criminal loser pals, was the day I realized I had lost all the love that I used to feel for him. In the last year, all he ever did was beat me down and hold me back. Now that he's gone, I can start a new life."

"And I'm sure you'll do just great, Darlene. What do you say we go for a soft drink and something to eat; are you hungry?"

"Now that you mentioned it, yes. I haven't had a thing to eat all day."

"Any place special?"

"How about, *The Kitchen Abode*. It's within walking distance and the food is good."

"That sounds fine to me," I said.

I helped her with her coat and we walked to the restaurant for a nice dinner plate special. As we walked down the street, we held each other's hand and talked. We talked a little more about Vern and his bullshit, but I moved the conversation to a new direction.

"When are you going to get a new phone put in?" I asked.

"Oh, I've already taken care of that, someone will be out in a day or two."

"Are you going to give me your new number? I said, half jokingly and half serious.

"Well that all depends," she smiled as we continued to walk.

"And that would be?" I asked.

"If you call me of course, silly," she answered, and then patted my butt.

When we sat down in the restaurant, she gave me her new phone number. We ate and talk.

"What day is it?" she asked spontaneously.

"Today's Wednesday, why do you ask?" I said.

"Because I'm planning on going back to work this coming Monday."

"Hey, that sounds encouraging. You're getting your life back on track," I said.

"I called my boss and he said I've got plenty of work to do."

Then, and from right out of nowhere, she asked the ugly question again, "Christopher . . . what do you *really* do for a living?"

"I'll be honest with you Darlene. Right now I'm in between jobs, but still trying to write. Sometimes I'll do a little work for my friend, Billy; he owns a party store."

"Surely you don't make that much; where do you get the rest?"

"Well, if you insist, I borrowed it," I replied with a straight face.

She didn't seem to react in any specific way, but instead, simply turned to a different conversation-probably not wanting to embarrass me. We did a little more small talk in the restaurant, and then walked back to her place.

"Would you like to come in for awhile, maybe a little after dinner drink?" she asked shyly. I glanced at my watch; it was after eight and I told her I had to go. I added that I had some important things to do, but I'd call her in a few days. We kissed each other and said our good-byes, and then I walk to the nearest intersection and waited for a cab.

When I entered my apartment, I took a slow look around and realized just how lousy my place really was. I knew that I was living a lie. My drinking was out of control and so was my mouth.

There was about two-thirds of a bottle of scotch on the table and some cold beers in the refrigerator. I grabbed a beer and sat at the table, then poured myself some scotch. I sat by the window, as I always do, staring out and looking down. My eyes were not focused at anything in reality, but instead, were gazing into the future. If I were to get involved in Darlene's life, then I would have to do something different with mine. What might that be? I hadn't the slightest idea. I sat at the table that night drinking, smoking and thinking. By four in the morning, I laid down to sleep.

Mr. Sandman woke me up the next day crying like a newborn baby. I had another reoccurring nightmare. I couldn't remember if I was screaming and yelling during my dream, but for some reason, I woke up crying instead.

I took my aspirins and sat at the table and waited for the sweat to dry and the shakes to go away.

After about an hour and six cups of coffee, I was feeling better, and during this time, I added some new details in my notebook. After entering my new notes, I put the book away and headed for the bathroom.

When I finished my bathroom routine, I put on some casual clothes and then counted my money. There was still the $9,800 in my bedside drawer and about $850 in my wallet.

Although I usually only check my mail about once or twice a week, I had discovered that my government check for $300 had come, and decided I'd cash it today.

I walked to the bank and cashed my measly check. Now I had $1,150 stuffed in my wallet. On my way home, I stopped to call Darlene just to see if her phone was working yet. No connection. I'll try again tomorrow. God, how I hate these dirty, nasty public phones.

I changed my mind about going home and flagged down a cab to visit Billy at the hospital. When I went to the information desk, I was told that the doctors had discharged him earlier this morning. I took another cab to his store where he lived on the second floor of the old store.

The place was closed, but that really didn't surprise me. I went around the side of the building to the entrance of his second story home. Because some parts of the rails were missing, I cautiously walked up an open stairway, which led to his porch landing, I knocked and his wife, Betty answered. She opened the door after peeping through the blinds.

"Hello, Sandman. You're here to see Billy, I hope?"

"Yes, Betty, is he awake? Otherwise, I don't want to disturb him."

"He's awake, Sandman. He's watching TV in the living room. Go on in, he'll be happy to see you."

I walked into his living room where he was inclined in a chair watching TV. I tapped him on the shoulder and he turned his head around to see. "How are you doing Billy? I read in the paper that you went to the hospital with chest pains. I went to the hospital to see you, and they told me you left, so I came here. God forbid, Billy, but did you have a heart attack?"

"No. It just turned out that my gastrointestinal track decided to switch places with my heart. Jesus, Sandman, the whole incident put me into a state of instant insanity. I didn't know what the fuck I was saying most of the time. I never stuttered, spit-up and threw-up so much in all my life. I damn near fainted about three times. I was seeing quad-triple vision. I told the police what you said to say, but I'd swear that most of my words were dribble. I was told by a nurses' aide, that the police eventually did bring tracking dogs, but they weren't any help. I've got to commend you, Sandman; it appears that you pulled it off. How you thought of that scheme so fast is beyond me; and why didn't you tell me you carried a fucking gun!"

"Billy, when was the last time you've seen where I live. It's a damn war zone down there. I wouldn't go out at night without a gun-I'd be an easy target. Believe me, if you're out in the city at night, you just may find yourself dead without one."

"Well, why in the hell would you go out at night anyhow? What is there to do in the middle of the night? What are you? A vampire? A thief or a junkie?"

"No, you know me better then that. Please, try and understand Billy. I just carry it for protection. I don't go around shooting people. But last night I had no other choice."

"How come I didn't hear your gun go off at the store?" Billy asked.

"It's got a silencer on it."

"A what on it!" Billy replied abruptly.

"A silencer. You know. A baffle screwed on the end of the nozzle so the gun doesn't go bang-bang real loud."

"Yes, yes, yes, I know what the damn thing does! But why do you need it?"

"Because, I'm fucking smart! Think about it. If I didn't have a silencer on my gun when you were being robbed, it would've most likely escalated into a gun battle after one of those guys heard my first shot. The lucky guy would have immediately started shooting back, maybe even killing you. I might have saved your life, Billy . . . oh, and don't forget, I put my ass on the line too."

"Okay, Sandman. I'm not as mad at you. After thinking about it, I was glad you were there and don't get me wrong, I was also worried for you. But what's done is done. Give it about a week or two and the whole damn thing will be old news. Let's forget it."

"When are you going to open the store again?"

"I don't know, Sandman, maybe in a month or two. I think my insurance is covered for this type of thing-if it isn't, I'm up shit creek. The lost income would really hurt me."

"I'm sure it'll all work out, are you okay for now?"

"Yeah, I suppose so; I'm okay, Sandman."

"Are you sure, Billy, you don't look like you're going to be okay?"

"I'll be okay . . . all right? Now, if you don't mind, I'd like to be alone. I'm not really up for conversation anymore Sandman, I just want to watch TV and hide from the world."

As I was turning around to walk away, I heard his echo from a wall, "Thanks for stopping by."

"No problem. I had to see if you were okay and I'm glad to see that you're well. You take care, and I'll see you on a better day, Billy."

"Sure," he replied without turning around.

After I left Billy's home, I wasn't feeling so sure about this whole incident. But I could breathe a little easier now, knowing what Billy told the police, even though I knew he was mad at me. One my way home, I had the taxi driver stop so I could pick up a couple of fifths and a twelve pack of beer. After I got home, I made myself a bowl of soup with crackers; there really wasn't much to eat. I knew that my groceries were low; a couple of cans of food, some dry bread, peanut butter, cold beer and an out-dated package of meat in the freezer. I would have to stock up again. After I finished my soup, I grabbed a dirty glass from the sink, a bottle of scotch and a beer. It's time for Sandman to sit down at the table, drink and watch the streets again.

The curtains were closed, so I slid them apart and opened another new daily perspective on living in the city. The time was about seven o'clock, and although I couldn't actually see the sun setting, I could see a very nice display of the clouds in a collage of pinks and crimsons, mixed with a little gray. I knew that there was a beautiful sun going down and disappearing between the buildings on the west side of town.

There seemed to be more traffic than usual tonight. Soon the dark will come and another night will bring out a different kind of humanity.

Some of them will be troubled, angry and hateful mixed with others who are fair, kind and happy; it's all a deadly combination.

I sat at my table just looking at the streets and wondering if anything interesting would play itself out for me tonight. It was warmer than usual that night-especially for this time of the year. Spring had come only about a week ago and I had my window opened without a screen. The breeze that came though the window this evening was unusual-unusual in the fact that the air smelt fresh and clean. I considered the idea of taking my beer and cigarettes downstairs and sitting outside on the steps. I've done that a time or two, but sometimes, goofy people walk by and stop, wanting to chat. Normally, that really doesn't bother me, but there are those occasions when a person is just too psychotic to talk to, and I hate that. It puts me in a position of either telling the person to go away, or simply get up and go back inside. And there are other times when I can feel that the streets are just too dangerous to be sitting on some outside steps with a beer and a lit cigarette. I decided that tonight, it might be pleasant to sit outside.

CHAPTER EIGHT

I wrapped my scotch in a paper bag, grabbed a few cold beers and headed to the porch steps to watch the traffic go by. It was a relaxing feeling, watching the cars drive up and down the streets. Everybody was going somewhere in the early spring warmth of a summer-like night. From time to time people would stroll on by, not stopping, but just saying, "Hi". This was the kind of evening that I really liked. To just kickback, and relax while watching the cars and people go by. Every now and then, I'd take a swig from my paper bag and a drink of beer, and then follow that by a long and relaxing drag on my cigarette.

I was enjoying the exceptionally warm evening on the steps when unexpectedly; I felt a cool breeze pass by. The wind started to pick up and a chill replaced the warmth. Well, there goes the fun, I thought, because I knew from the feel of a cold front passing through, that there would be a storm coming in. Off in the west, I noticed that there were flashes of lightning in the far distant sky. Within seconds, a light sprinkle of rain began to fall. It was time to go upstairs and watch the storm roll through by the side of my window. Just as I had made it through the doorway, the rain started to pour heavy. I sat at my table and turned the radio on. I could hear the crackles and pops from the static of the approaching thunderstorm.

A news lady interrupted the music and announced severe thunderstorms with lightning and possible hail, and wind gusts up to 60 mph or more. A few minutes after the announcement, I heard a powerful crackle of static in the radio. Suddenly, while looking out the window, I saw a thick bolt of lightning come down from the sky to meet its opposite target on the ground in the parking lot across the street. It lit up the outside brighter than day. It horrified me so much that it made me jump to my feet. The thunder followed immediately. It created a sonic boom so loud that it literally shook the building and rattled my window. For the first fraction of a second, I saw the thick streak of lightning actually dissipating right in front of me. The flash didn't just suddenly disappear, but instead, seemed to melt away in segments. I've never seen lightning this awesome, strike so close. When the lightning struck its target-a car in the parking lot-it caused it to explode in flames. I watched in horror as a man jumped out of the flaming car and started running from the parking lot on fire. He made it to about the front gate when he dropped and fell on his face.

I got up, and ran out the door to the parking lot. When I got there, the security guard had put him out with his coat. The man was lying on his stomach and when I turned him over I was praying it wasn't Scott. I was relieved to see that it wasn't him.

I put my ear to the man's chest to listen for a heartbeat, and then tried to give mouth-to-mouth and beat on his chest. After four or five minutes, I knew he was dead. The rain was cold and coming down hard still followed by the lightning and thunder that was all around. The smell of the wet, burnt, flesh-fumes smoldering off the dead man's body was more than I could take. It was putrid and it made a pasty taste in my mouth. The security guard came over and told me that he had called the police, and then asked me if the man was dead. I just nodded my head yes as I glanced at the car again. It was still on fire. As usual, I don't like to get involved, so I ran away in a different direction that would eventually lead me back to the safety of my apartment. I had to wonder what horrible crime this man did to God to strike him so wickedly dead.

As I grabbed the doorknob to my apartment, it suddenly occurred to me that I had left my door open when I ran outside. I started to panic. Soaking wet, I ran to my bedside drawer to see if my money and gun were still there. To my horror and disbelief, they were gone. I dropped to my knees and started to run my hands through my wet hair while I laid my chin on my chest, not believing how stupid I could be. After a couple of minutes, I slowly stood up and looked around. My empty holster hung on the clothes rack and my fifth of scotch still sat on the table. I walked to the door and closed it, this time locking it as well. I sat down at the table and poured myself another glass of alcohol.

I drank while looking out the window watching the parking lot. The police, ambulance and fire truck were busy doing their jobs, and all the while I sat quietly cursing a man named Scott.

I was quite sure that Scott didn't steal my money and gun, but I couldn't help but to be mad at him. I was afraid that it was he out there, lying facedown on fire. However, not all was lost, I still had $1,150 inside my wallet. But what really pissed me off was my stolen gun. The serial numbers had been filed off; so tracing it would be impossible. Damn! I thought to myself, it was a good straight shooter and hard to replace. But on top of that, there was the $9,800 I had lost. The money would be missed, but as a shrewd soldier, I knew I would manage to come back even stronger.

I finished my scotch that was still wrapped in a paper bag, with a few more beers. I sat and watched the thunderstorm roll through. After the last small sprinkle, and soundless lightning, I took my moist clothes off and laid down to sleep. But, because I was robbed, and my stupidity, my mind started racing with thoughts of retaliation, destruction, evil, hate, and anger. I had to put these thoughts and feelings away because I knew better than to try and fall asleep visualizing about getting-even schemes. Come the morning, it would be all Mr. Sandman could handle, fighting his nightmares and the maelstrom that always follow. I chose to think about Darlene, and within a short time, I fell asleep.

"I'll kill you, I swear I will you rotten motherfucker! I'm going to scrape your guts out with my bare hands and eat your bloody heart raw! I'm going to . . ."

"Shut the fuck up in there you stupid bastards! You woke the whole goddamn floor up! I swear I'll call the fucking cops if that fighting doesn't stop. Take it outside ass-holes."

The banging on the door woke me up and hushed my screaming. I was digging at my sheets, sweating and bleeding where my head hit the wall, and I was shaking badly. Not surprisingly, I had a terrible headache, but the first thing I did was get rid of that son of a bitch banging on my door.

"Okay, mister," I placated him, "just a little misunderstanding. I'm sorry. It's all over with now."

"Damn well better be!" he yelp, as I heard him slowly walk away.

I moved faster then usual, taking my aspirins and going to the bathroom. That was a peculiar nightmare. Strange, I thought, because this was no reoccurring nightmare. It was totally new. I dreamt that I found the man who had stolen my gun and money. I was beating the life out of him and to me, this felt like a more normal dream, because it was some other man, not me, who was being tormented. It might even be considered the type of dream that anyone might have after being robbed. I grabbed my notebook. This one was definitely going under my new title: Normal Dreams.

It was nine in the morning and after recuperating from my dream, I felt better and more normal for a change. I showered, and then fixed myself a can of corn beef hash, which I ate while listening to the late morning news. I decided today to give Darlene another call after I was up and around and ready to face the day. After putting my mess away, I put my nice clothes on, turned the radio off, and headed for the door. I closed and locked it behind me this time, as I *almost* usually do. I walked about half a block before I came to a public phone booth. I hate to use these damn things-they're old, out of date and nasty.

Inside the booth was the odor of at least a thousand different people. The plastic windows were covered with scratches, graffiti, drools of spit and other substances, which I decided not to try to identify. The shapes of cars and people passing by made odd wiggling motions through the concaved plastic.

But the public phone itself is an abomination of invention. There's no sanitation. And who knows what kind of people handled the receiver and pushed all the dial buttons. They always feel sticky or wet. Every time I use a public phone, I think of it as an unsanitary designed piece of technology. God only knows what type of germs and bacteria could be transferring on these phones-maybe even in a terrorist form as a worst-case scenario. I put the money in and dialed Darlene's number.

"Hello?" she said after one ring.

"Hi, Darlene, it's me, Christopher."

"Well hello, honey, it's nice to hear your voice again. I miss you. Are you calling to come over?"

"You must have ESP; you wouldn't mind?"

"Of course not, but could you make it for about one o'clock. That would give me time to pick up my mess and freshen up."

"Okay, I'll see you at one," I said and then dropped the phone and set myself free of the filth from that nasty cubicle. I walked a block down to a corner coffee and donut café, ordered a coffee and snatched a stray ashtray.

For some strange reason, I began thinking about phone booths of all things. An idea came to me that I thought would resolve the ugly situation, but it would be an added tax burden. I thought, how could the phone companies make their public phones more sanitary? My idea was that if they could make their phones water and corrosion proof, they could design their booths with a sewer drain and a garbage-type disposable built inside. With a showerhead installed on top, once or twice a week, the phone companies could spray down the inside with cleanser containing antibacterial solutions and disinfectants that kill harmful bio and chemical compounds, followed with a final spray of water. All the shit would wash down the drain and after some minutes, the booth, including the phone would be clean and dry. Ah fuck it, I thought to myself, these things are a dying breed anyway. People carry cell phones today, but I still thought it was a clever idea just the same. Thinking of the time now, I glanced at my watch and discovered that it was quarter to one-time to go see Darlene.

I flagged down a taxi and within a minute or two, and I was on my way to her apartment. When I arrived at her front door, I noticed she had a doorbell that I hadn't noticed before. I pushed the button and Darlene opened the door. As always, she looked gorgeous. Her attractiveness is almost impossible to describe with words.

She asked me in, then closed the door, turned around and gave me a big hug and a kiss. I felt her breasts pressed up against my chest, and I grew an erection that I knew she could feel. I felt embarrassed, but she simply told me she was flattered.

"Wow, Christopher. I've never had a hug and a kiss like that before. I'll have to sit down and catch my breath." I sat down next to her and she laid her head on my shoulder. I couldn't help but to stroke her soft hair. After a couple minutes, she slowly stood up.

"So how have you been?" she asked.

"Well, things could be better on my home turf, but right now, I'm feeling quite nice."

"Good. I made some fresh coffee, would you care for a cup?"

"I just left a coffee shop, but a real fresh cup sounds good. But first, may I use your bathroom?"

"Sure, I've got two, take your pick."

I used the one from the hallway and emptied my bladder of the two cups I drank at the coffee shop. Her bathroom was so clean that I felt I must be very careful not to piss on the toilet rim, or even worse, on the floor. Before I washed my hands, I made sure to put the seat down; a little secret a woman told me long ago. When I stepped out, she had prepared a nice coffee set in the dinning room. I sat at the table across from her.

"So, Darlene, you'll be going back to work on Monday?"

"Yes, and I'm really looking forward to it."

"That's nice. That's good. Listen, Darlene, I hate to bring up this conversation, but I'd like to know, how did everything turn out with Vern's body?"

"It's all over with. After all the politics, the police said I could give his body to the university and so I did. They thanked me for the corpse and I haven't heard anything since. I absolutely love the freedom. But please, don't think I'm morbid?"

"No . . . oh no, I don't. I just hope that you're getting along well. I'll never bring the subject up again. All the bad is now in the past and I'm looking forward to seeing you grow even more happy and cheerful."

"Right now, Christopher," she said with a healthy smile, "I'm the happiest I've been in years."

"Then that makes me happy too."

"Thanks, Christopher, you have such an enjoyable outlook on life, I don't know where you get it? . . . Oh, I'm sorry, you're out of coffee, would you like a refill?"

"Sure."

"And I know how you like your coffee too. A half level teaspoon of sugar with a splash of milk."

"You got it right babe," I said with a touch of affection.

"Is there anything else I can get for you?"

"No thank you, I'm fine, but I would like to talk to you about something that's been on my mind?"

Her eyes presented a kind of wonder-worry look. Without question, she got up and sat beside me. "Is there something wrong?" she said in a concerned voice.

"I don't know, Darlene, you'll have to tell me. I have to ask you a question and I pray you'll be honest. Okay . . . here goes; do you feel any physical attraction for me?"

She sat and thought for a few moments, and then took my hand. "Let's we go sit on the couch?"

As I stood up and walked to the couch, she walked right beside me with her hand still in mine. She gently sat me down and removed her hand while sitting next to me. There was a strong sadness in me, (the wheels have been turned around again), and I felt that rejection once before.

I remember long ago, I fell in love with a girl who evaded my every affection no matter how hard I tried to gain her attention. The pain in my heart that I had to endure was the worst pain that I could ever remember. No matter how hard I tried to please her, she only placated her affection to me . . . and now I was ready to feel it again.

"Christopher, the first day I saw your face, you lit up my heart. The infatuation was overpowering. I can't explain it, but you're the most handsome and attractive man I've ever met. I've never believed in love at first sight, but for some reason, I knew I had to see you again; that's why I told you where I worked."

"Did you mean it when you told me you loved me?" I asked.

She softly caressed my face and hair, then smoothly rubbed my chest and waist and gently put her hands between my legs. She then slowly put her arms around my neck and hugged me while whispering the words I so wanted to hear: "I really, really do love you, Christopher." I hugged her back and kissed her lips with a passion I've never experienced before. After our lips parted, I could smell her soft breath, and that essence sexually aroused me. "I'm in love with you too, my enchanted Darlene; we'll outlive the stars."

I had a smile on my face and my heart was filled with joy; more joy than I've ever known. I also felt blessed, because she . . . so beautiful beyond words was the first woman who ever-ever loved me in return. We held onto each other as if we were afraid the feeling would steal itself away. We kissed deeper, but gently, and her very being was like magic beyond my physical realm. My mind was cleansed of all the lost and lonely feelings, and replaced with the warmth of Darlene's body and the love she was giving up for me. She suddenly stood up, as if by some command, and faced me with her outstretched arms while saying, "Please take my hand."

She led me to her bedroom and we lay on her bed. We made love to each other and I knew from this day forward, a memory would be engraved, and saved forever inside my head. I said a simple secret prayer that I would never lose her, and that she would be mine until the stars do indeed, stop to shine.

After the sexual post-play, we started behaving normal again. We basically made small talk, while sharing some drinks. I didn't stay very long. She told me she loved me again and I echoed her words. Then she told me she wanted to be alone, and I did as she asked. Before I left, she assured me that today was the best day of her life, and she also assured me that she'd never forget.

"Darlene?" I asked feeling a little uncertain.

"Yes?" she replied with glistering green eyes, looking up and directly into mine.

"Are we really going to see each other again? Or was this the culmination of something that was beautiful, but now ending?"

"Oh . . . Of course not, my dear Christopher. I wouldn't have told you that I loved you if I were to never see you again," she said smiling and obviously thankful and relieved for the added spare time.

"Listen, Darlene, I just wanted you to know, I have no intention of keeping you on a leash. I want to give you your space, and the room you need to grow and pursue your career and dreams. All I want is to be beside you sometimes."

Again, when I entered my room, I realized how small it was compared to Darlene's apartment. I changed my clothes into some casual rags, and then sat down at my table and started to drink. I had enough scotch and beer to keep me awake and pacified for the rest of my disgusting night. I glanced out the window and stared at nothing. I needed to re-think my future. Darlene was right. Starting a relationship was something you had to think about.

And as I was thinking about the future, I found myself in the past. I sat and thought about my childhood and the place where I was raised. There was no love lost and no fine thoughts at that damned military base. I left as a First Lieutenant, and entered the streets as a bum. I also wondered where Scott was and how he was getting along. I sure did miss the old son of a bitch, and I missed his conversations too.

He said in his note that he would be back later, but I had to speculate that maybe by now, he was dead. Then I thought about being robbed. And the more I drank, the madder I became at the thought of losing my gun. I decided to take a walk later after dark, a walk in the night to find a new one.

I kept drinking until it was almost midnight, when I was buzzed and brave enough to go out on the streets and get a new gun. I left all my money in the drawer with my wallet-$1,150. I closed and locked the door behind me. It was time for a walk in the dark.

I walked down Steward Street heading further east where the bad boys hang out, and the prostitutes talk to horny old men from the side of their car windows.

I carried a knife taped to the inside of my leg.

About eight blocks down, I saw the juvenile I was looking for. Not someone I knew, but rather someone I needed. The punk was leaning with his back on an abandoned, boarded-up building with one foot propped up against the wall and the other on the ground so he wouldn't fall. As I walked up toward him, he simply looked me up and down showing absolutely no fear. I stopped to talk.

"What're selling?" I asked.

"What do you think I am, a fucking street vender?" he said with a cig between his lips.

"No problem. Sorry I bothered you," I said and started to walk away.

"I've never seen any cops like you on this side of town before," he said trying to prompt me.

I stopped and turned around, "I'm not a cop. You must be new on the street." I teased him.

"How should I know you're not a cop?" he replied.

"You don't. But that's the chance you take. Besides, you're too young anyway."

"What do you want?" he spoke abruptly.

"I'm looking to buy a gun with a silencer," I said.

"Hey, Stud, Mellow, come out here and see this one," he said as two other punks came out from around the building.

"This dude wants to buy a gun with a silencer on it. Do you think we can help him?"

"Why don't we just jack him and take his money," Stud replied.

"Don't be stupid, asshole. It's not good for business. Why do you always want to rob or kill the customers?—You ignorant shit. Now look, fuck-wads, I'll ask you two dick-heads again-can we, or can't we, help this man?"

"Yeah, I know what he's looking for. I can go get it and be back in thirty minutes," Mellow said, while standing and waiting for his master's command.

"How much is this going to cost me?" I asked.

"You can have it for eight-hundred. Go get your money and be back in half an hour. Mellow?—Go get the gun."

I turned around and casually walked west for a block, turned a corner and took a few more steps. After I knew I was out of his sight, I started running north and then west down the side streets. I zigzagged around until I reached my apartment and took out eight hundred from my wallet. I took off running through different side streets, and then slowed my stride down when I saw the guy still leaning against the wall. It had taken me about thirty minutes to make the round trip, but I didn't see the other two guys from where I stood. I cautiously walked up to the guy against the wall.

"You don't look as if you've moved an inch," I said.

He ignored my trivial comment and asked, "Did you bring the money?"

"Yes," I answered simply.

"Mellow?—Get your ass out here," the master demanded.

Mellow stepped out from the corner of the building pointing the gun at me sideways with a big, wise-ass stupid grin. I glared at him straight in the eyes, returning evil for evil and calling his bluff, because only a fool would sell a loaded gun.

"I see you found what I'm looking for," I said hoping the gun was indeed unloaded.

"Mellow! Quit pointing that goddamn gun and just give it to the man. Stud—If this guy tries anything stupid . . . kill him," the master punk said nonchalantly as he put his foot down.

Mellow handed me the gun while I was handing over the money to the leader of Mellow and Stud. It only took me less then a second to see that this gun was mine-my stolen gun. "Where did you get this gun?" I asked, before realizing what a dumb question that was.

"I don't tell, I just sale. Now get the fuck out of my sight," the leader replied.

I started walking home taking the route I took before. After I was out of their sight again, I took off in a mad dash just in case they wanted it back.

When I arrived home, I put the gun back in my bedside drawer along with the last of my money-$350. By now, it was going on two-thirty in the morning. I went to the refrigerator for a beer and sat back down at the table with my scotch and cigarettes. I reached over and turned the radio on to some talk show. It was Saturday night and I had already put my ass on the line again, so I decided to sit and drink through the rest of the night until I saw the rising sun. But my eyes didn't see the sun rise; I was too tired to keep my peepers open and before six in the morning I was sleeping-and Mr. Sandman would come again.

I woke up Sunday at two o'clock in the afternoon. After a big yawn, I stretched out my arms slowly waking up. I got out of bed and washed my face, and then brewed some coffee to help me awake. I sat at my table drinking a cup and smoking a cigarette while I looked down at the bed. Something didn't feel right. Then it struck me like a knife in the heart. I woke up with no nightmares, and I remembered my waking dream! I dreamt that I had the power of levitation and could hover and fly as low or as high as my desires. My God, what a wonderful dream it was. It made a very deep and mystical impression inside of me. I felt angelic, and even envied by the people on the ground that were watching. I've heard about people having these kinds of dreams, but never in my life did I think it would happen to me. I had to get my notebook and record this down; again, under the title: Normal Dreams.

It was a bright and sunny day with a warm breeze sifting softly through my unscreened window. Again, the air smelt fresh and clean on this pleasant, Sunday afternoon. The sky was a lovely blue, almost translucent, which I've never seemed to recognize prior to this day. It put a smile on my face, which reflected my attitude about the dream I had. I felt in a good mood. The more I thought about it, the more exhilarated I became. For the very first time, and in five years, I felt normal again.

CHAPTER NINE

Suddenly, I was bewildered when I heard a knocking on my door. My first thought was that it was Darlene coming to make an unexpected visit until I heard a man's voice resonating from the other side. I approached the door with uneasiness.

"Who's there?" I asked innocently.

"The FBI, open your door," a deep voice said.

I wasn't quite sure what to do. The knocking persisted and the man's voice continued to tell me to open my door. I was busted!

They finally caught me, and I had no other options but to do as they told me to do. I cracked the door ajar.

"Yeah?" I asked, looking stupid and wearing wrinkled clothes.

"May we come in?" one of them said.

"Can you show me some identification?" I asked. They held out their I.D. and badges.

"Now . . . may we come in?" he asked again.

"I suppose so," I said and opened the door.

They walked in like soldiers wearing black suites and shinny shoes. "We'd like to ask you a few questions, if you don't mind, of course?" the taller one said in his deep voice again.

"Shoot. I've got all day, but if you're going to take me away, I want to see a warrant."

"Do you go by the name of, Mr. Sandman?" the shorter man asked.

"Sometimes, but what's that got to do with you?" I asked quizzically.

"Just answer our questions," he said.

"Do you know a man by the name of Billy Richers?"

"Yes. I buy my booze and cigarettes from his store-why?"

"Do you know of a woman by the name of Darlene Osmon?"

"Yes, I do, now I want to know what this is all about." I said in a slightly louder and demanding tone.

"Please, sir, just answer our questions," the taller man said this time.

"Did you know Darlene's husband, Vern?" the taller man spoke again.

"No, I didn't," I demanded.

"Did you know of an attempted robbery at Billy Richers store last Friday night?" he asked.

"Yes. I read about it in the papers," I replied while trying to stay calm.

"Did you know a vagrant by the name of Scott Sellers?"

"Yes. We became friends until he left," I simply said.

"Okay, Mr. Sandman, please come with us," the shorter man said in a matter of fact tone.

"Hold it here, mister! What do you mean . . . go with you? Am I under arrest?" I requested.

"Just come with us, sir," the tall man said while grabbing my arm.

"Bullshit! I want to see a warrant!" I yelled while resisting his hold on me.

The taller man twisted my arm to the point of breaking and said, "Look, sir, we can do this thing two different ways. I'd suggest you do it our way."

"Do what thing! Who the hell are you two? Where are you taking me? I demand an answer!" I shouted like one of my victims.

"Look, mister, we're simply going to walk out the door, shut it and lock it. Then we'll get in a car and go for a little ride and everything will be just fine," the shorter man replied.

I decided to do it their way. I was busted. They knew who killed those two men, and so did I. When we got outside, I looked for a police car but didn't see one. Then it occurred to me that these men were FBI agents; they don't drive around in police cars.

I looked around and saw a couple of suitable sedans parked by the curb. They were both strongly holding my arms when I noticed the shorter man wave with his first two fingers at someone. To my disbelief, a long black limousine pulled up in front of us. The shorter man opened the door and the tall man calmly ushered me in.

We sat down in different seats when I noticed that this was not an ordinary limousine. There were all kinds of creature comforts in this car, and to my amazement, even a sparking toilet seat with curtains.

We started to drive off as I watched the scenery move by through the dark, stained, and hidden windows. "Would you care for a glass of scotch?" the shorter man asked.

"Yeah . . . sure, but without ice, and make it a double," I said poignantly.

He reached over to a cabinet and opened two small doors. Now, I was sure that these guys were not the police. He poured me a glass of scotch without ice.

"I think you'll find this blend much to your liking. Would you care for some music?" the shorter man added.

"Sure. How about some good rock & roll?." I asked like a smart-ass.

To my surprise, he turned a CD on. I didn't know who theses two were, but I quickly surmised they were definitely not your typical, FBI.

The scotch was good as was the music. The scenery was getting better as we traveled west on Steward Street. I even saw Darlene's apartment as we drove by.

I watched through the windows, as the homes got bigger and better. After about thirty minutes or so, we had made numerous turns until we were in the country with rolling hills and approaching mountains. Although I was acting unafraid, I was becoming concerned. The farther we drove away, the closer I thought of Darlene. But I stayed steady fast with my situation.

"You guys aren't taking me to any authorities . . . are you?" I asked while sipping on my scotch and listening to the music. Both of them sat silent. "Say, you guys wouldn't have a cold beer . . . would you?" I said, still wearing the wrinkled rags I slept in. He handed me a can of beer and asked if I'd like a glass. I shook my head no. I happened to notice that there was a console beside me with recessed holes and a big ashtray. I sat my glass of fine scotch and beer in the holes. There was a good song playing, and so I listened while we drove along. I was hoping not to have a panic attack.

"You guys like this music?" I asked.

They didn't respond. At about the same time, a phone rang. The shorter man answered it and the volume to the music automatically went low.

"Yes. No problem, sir. Fifteen minutes. Right," he said before he hung up.

"That was your boss, wasn't it?" I asked as the volume went up. Again, there was no response. "Say, any of you two guys have a cigarette?"

"Look again at your side in the stand," the tall one replied. I looked and discovered an unopened pack setting in a slot with a push-button lighter next to it. "Cool. You guys have everything in this limo," I said as I lit a cigarette. They glanced over at me and then to each other. "Hey, you guys know any good jokes?" I asked, faking fancy for fear.

The tall one turned his head and temporarily looked at me. At this time, I definitely knew they were not the FBI, but at the same time, I was wishing they were. I thought that they might be some kind of higher echelon of thugs-gangsters. Who ever they might be I hadn't a clue; and I still didn't know where they were taking me to.

For just a split second, I panicked and my sight turned to tunnel vision, and then back to normal. I did my best to compose myself, but that's very hard when something unexpected and possibly dangerous is happening to you. Were they eventually going to kill me? I honesty didn't know. My thoughts and mood turned more serious as I wiped some faint sweat from my forehead.

"Come on you guy's, give me a clue, where are you taking me to?"

They didn't say anything. The only thing I could do was just sit back, listen to the music and watch the mountains get bigger as I drank my scotch and beer and puffed on my cigarette.

My ears started to pop as we ascended into the mountains. There were no more buildings or homes to look at until we made a turn. We were headed toward a large multi-brown structure, nestled between the peaks of two mountains. As we got closer, we approached a road and stopped. It had big iron brown gates that opened up into a camouflaged driveway surrounded by tall, steel-green fences with razor wire on top. The music went dead and the phone rang again. The shorter man answered and whispered. I couldn't hear what he said.

As we approached the gates, they opened inward, obviously expecting us. When we drove through, I swiveled my head to the right and left. There were two military men standing at the ends of the gates holding Ak-47s. I gulped down my scotch because my surroundings were telling me that I would be getting out soon. The long driveway ended at a big circle turn but we stopped when the limo was parallel to the front of a fantastic looking building. As we got out, I noticed that there were cameras surveying every inch of ground, and there were even more soldiers protecting the building. The agents escorted me to the front doors with no knobs.

The doors opened and with a slight push from behind, they followed me in.

"Please, Mr. Sandman, follow us," the tall one said.

I saw more cameras as I turned to follow them. The foyer was large and the ceiling was high. If I hadn't seen the outside, I would have sworn we were in an exclusive hotel. We walked to an elevator door and the short man pushed a button indicating up. It took only a second before the elevator door opened, and we stepped inside.

I saw the short man push a button with the number 4. The schemes of the buttons went above *and* below the first floor. We ascended fast and then the door slid open, revealing the splendor of someone's home.

"Please come with us," the short man spoke. I followed them through a wide hallway, which eventually led to a large living room and a view that some people would kill for. The furniture was stupendously plush and the walls were curved leading up to a huge vaulted ceiling that gave the illusion of never ending. "Please take a seat here," he said while pointing to a chair. I sat down and noticed that there were no cameras watching us. I sat waiting for whatever was to happen next when I saw a man walk in. He had white groomed hair that was slick backed on his head, a neatly white trimmed beard and he was dressed very expensively. He started to talk to me.

"Hello, Mr. Sandman. It's nice to see you again, my old friend."

"Scott!" I yelled.

"Are you surprised? I'm very sorry I've inconvenienced you, Mr. Sandman, but there's something very important we must talk about."

"I thought you were a bum!" I exclaimed.

"Or so you believed, but as you can see, the friend you made on that cold February night, was not really me."

"Then who are you, Scott, and why am I here?"

"We'll talk about it later after I show you your chambers. You'll need some time to clean up and get out of those rags."

I got up and stood by the foot of the chair until Scott motioned for me to follow, and as we walked, I talked. "What's going on Scott? That trip was almost more than I could bare. I wished I knew you were behind this, and by the way, whose palace is this?

"Well . . . mine, and a few others."

Listen, Scott, I'm not sure what this is exactly leading to, but I don't like the smell of it, this whole damn thing is starting to really bring back bad memories, and I don't like it."

"Don't concern yourself; you're safe. As I mentioned earlier, I'll explain later. But, for the time being, just follow me."

I followed Scott down a hallway passing various doors until we stopped at one, which he swung open. "Here's your bedroom," he said waving an arm.

"Well now I'm really puzzled, Scott, it's as big as a gym."

"And it also has a full bathroom; why do you look so surprised?"

"I thought you were going to cage me in a cell with a shitter and a shower or something?—You said chambers?"

Scott laughed softly, "I know you're fucking with me. Here's your bedroom chambers; the place you'll sleep and bathe."

"Oh . . . I see," I mumbled while looking around.

"As a matter of fact, Sandman, this whole floor is yours. This will become your new home for awhile."

"Really?" I asked seriously.

"I'm afraid so." he replied simply.

"When are you going to tell me why I'm here, and what for?"

"After you shower and make yourself presentable."

"If you need anything, just push any one of the red buttons you see on the walls. Your clothes have been laid out on the bed-now go take a shower and get that foul smell off your ass. You have absolute privacy."

Scott turned away and walked out the door, closing it behind him. I walked through the bedroom and into a bathroom, which was bigger than my apartment. There were mirrors, bottles and boxes that lined the long counter with two sinks and a toilet that literally looked like a king's throne. The tub was huge and enclosed. There were water nozzles from my head to my toes. I undressed and stepped into the tub, and the transparent doors automatically closed.

I bent down to turn on the water when I discovered that there were no knobs. I stood back up and looked around the lining of the tub and was bemused until I saw a digital screen on the back of the bathtub wall. As I was studying it, I wasn't quite sure what to do. It had different features on it and as I read them I became amused.

There were buttons that controlled the temperature and pressure of the water. Some of the buttons controlled different water nozzles, and were labeled with numbers. I pushed a button, and at that moment, the water came on.

I was braced for the initial blast of cold water like other showers do, but instead was given an instantaneous burst of warm water that felt like a heated pool. The showerheads sprayed a fast pulse of water, which felt like millions of little fingers, massaging every small inch of my body. It was absolutely delightful. There was also a dispenser, which allowed me to select a variety of soaps, shampoos and conditioners.

When I was through washing my body, I simply had to touch the *off* button. That was the most fun I'd ever had taking a shower, and afterward, I pushed the *dry* button, which eliminated the use of towels. From the water nozzles, came jets of warm air that dried my body-it only took a minute or two. Standing at the vanity with the long and tall mirrors, I sampled different antiperspirant & deodorant sprays. Then I shaved my face, brushed my teeth and combed my hair. For the last few touches, I powered my body and splashed on some exquisite cologne. It smelt like fresh springtime air, combined with a small scent of a sanitary fragrance. It made me smell like a rich man.

I walked into the bedroom and looked at the clothes laid out for me. The shirt was white and labeled silk. The pants were deep blue and crafted from a fine fabric.

My underwear was silk and cotton and my socks were made of a medium thick nylon.

Beside the bed was a pair of shiny, black-leather, ankle boots. After I put everything on, I felt distinguished wearing the smoothest, and form fitting clothes I've ever worn. Not only did I smell rich, now I looked rich too. After one last glance in the mirror, I pushed one of the red buttons, and within seconds, there was a knock on my door. When I opened it, there was a military man standing at attention.

"How may I help you, sir?" he said.

"I would like to see Scott."

"Please make yourself at home in the living room," he answered, and then walked away.

My first instinct was to look out the wall of windows. They didn't open, but yet I could smell fresh air blowing in softly from somewhere. There was a nice view of the mountains, and a stream down below which turned into a waterfall, and then a mist of water, which created within itself, a small and brilliant rainbow. I watched for a while, and then sat in a chair where I noticed a remote control positioned on top of an end table. I picked it up and looked at the buttons. Nothing seemed unusual, I pushed the *on* button and a large portion of the wall moved sideways, revealing a big curved screen.

The sound from the pristine picture it displayed seemed to come from everywhere. However, I quickly turned it off when out of the corner of my eye, I saw the military man standing at a small distance next to my side. The large wall started to close, and I was compelled to watch until it was through, while the man watched patiently and silently too, and then he spoke. "Mr. Sellers is waiting for you in his guest room. Please follow me."

I followed the man through a set of steel doors, which led to a flight of stairs. We walked down one floor where he then led me to a new and extraordinary room. Scott was sitting in a chair smoking a cigarette, and sipping on a glass of scotch.

"Well, Sandman, you look nice. Please take a seat and make yourself comfortable. Is there anything I can get for you before we eat?"

"Sure, I'll have what you're having; scotch I'll bet."

"No, it's wine. Sergeant Cooper, please pour my friend a glass of scotch."

"Scott, I'm confused-who the hell are you? Tell me what in the hell is going on, and what am I here for?"

"Sandman, the night you first met me was a fluke, a stroke of luck on my part. I was personally investigating, and surveying you. I purposely dressed like a bum and walked your streets for days seeking you out. That was part of my scheme.

As I was seeking you out, and in time, I would have made the initial confrontation. But to my surprise, you chose to seek me out instead, and invite me into your apartment. You did my work for me. By the way, how did you like my pig farm, schoolboy story?"

"Yeah, real charming, Scott. You got me on that one. But you still haven't answered my question. What am I doing here, and while I'm at it . . . would you please answer my question and just tell me who the hell you are?"

"I'm your friend, Sandman, now calm down and let me explain. I'll have to answer your simple question in a rather complicated way. First: I'm rich. Second: I'm deeply involved within the government and military.

The building you're sitting in is embroiled in a government conspiracy cover-up. Beside myself, only a handful of women and men know the real identity about you. Security is of the utmost important. I have access to the CIA, FBI, all the military branches and other government agencies. Most people do as I say and keep their mouth shut, or they disappear from the face of the earth, as you may, if you don't cooperate. Nothing personal, Sandman, but I think you'll see in time, I'm making a new future. Call it: a new beginning."

"You're kidnapping me. I know what this is all about. I thought the bad dreams were all over, Scott. I thought it was just a matter of time before the doctors found a cure for my nightmares."

"And so they have, Sandman. How did you sleep last night?"

"I had a dream . . ."

"You had a nice dream, I trust. Would you like to tell me about it?"

"Not really. But you're right, Scott, it was a very nice dream."

"No more nightmares for you my friend, Sandman, I'm about to turn your life around. You'll thank me later. Now, I'll bet your hungry, let's have something to eat and we'll talk more." Scott stood up and proceeded to walk while I followed. He said nothing until we were standing in a dining room with another man.

"I want to introduce you to my associate, *Dr. Bradly Hues*. He'll be working with us."

"It's a pleasure to me you, Mr. Sandman," he said, extending his hand with a smile.

"Likewise," I replied as I shook his hand.

"Please, gentlemen, take a seat. Sandman, I want you to meet one of my prestigious chefs', and your maitre d', *Henry*."

Henry clicked his heals and bowed his head slightly, "At your service, sir."

"Likewise," I replied.

Henry left momentarily, and returned with a bottle of wine. He poured each of us a glass, and then disappeared through a set of swinging doors. I picked up my glass and took a slip. It wasn't exactly my cup of tea, but it tasted pretty good.

Scott began to speak, "Mr. Sandman, Bradly is rather new here, as a matter of fact, he just arrived. His professional career as a scientist is genetics, and he's going to be collaborating with me. He's been stamped and certified to work here. We're all quite safe, so please, talk freely.

Although Bradly is well aware about what we're doing here; code name: *The Genesis Protocol*, he knows very little about you, personally.

And so I thought to myself, who better to tell your story but you. Would you do the honors for me? And please, in your own words, start from the beginning."

"So you must be the head-man of this facility?" I asked Scott.

"That's correct, Mr. Sandman. Now please, you're avoiding my question."

"But I'm still in the same city?"

"Well, yes, relatively speaking-this is where you were cloned, but not raised. Now please, Sandman, tell Bradly your story," Scott answered.

"Okay, Scott . . . from the beginning," I said as I turned my head, "Bradly . . . I'm a clone. Cloned from the flesh of a dead man-an outlaw gunned down in the old west. My real name is my father's name: *Jeff Dunbar*, and I am the clone of this notorious outlaw. He was the leader of a gang in the mid to late 1800's. My father was dug up from a mound of the sand dunes at the base of the Blue Mountains, next to the Snake River Plains. The . . ."

Suddenly, Scott interjected, "Jeff Dunbar died in 1898 in a shoot-out with a bartender. His remains were still in quite remarkable condition when they put him in cryogenics in the late 1900's.

From that time on, his body mysteriously disappeared, only to end up in the hands of a few men and women scientists and billionaires, known as the *Society*. And that's where we are right now. This place is where the first ever cloned human was grown."

"*Interesting*," Bradly simply said, showing no emotion.

"Sorry to interrupt, just updating Bradly, continue with your story, Jeff," Scott gestured.

"Interesting? Did you hear that, Scott? I'm fucking *interesting*. I'm a goddamn miracle monster to you, Mr. Hues!"

"Please, Jeff, calm down. You're new to Bradly, he's never talked to a clone before."

"Well, tell Mr. Hues what a horror and hell my life has been, Scott. Go ahead, tell him yourself!"

"Okay, Jeff! Don't get so worked up, and sit back down. I'll let you tell us what a horror and hell your life's been. But remember, you're here because I wanted to right a wrong, and make your life a new and better one."

I sat back in my chair and relaxed. I looked at Bradly and then Scott. They were looking at me as if they were taking mental notes. After a few seconds, Scott pushed a button from the end of the table.

"Henry, would you please serve us our meal."

Henry walked out with a serving tray and quietly placed down our plates. He refilled our glasses and asked Scott if there would be anything more.

"Thank you, Henry, that'll be all for awhile. Now, gentlemen, let us eat and enjoy this wonderful food. I hope you like steak; I love it, but I can only have it twice a week. Don't forget your salad, it's full of important vitamins and fiber, and the dressing is fabulous. It's low cal and fat free. Please-enjoy," Scott said like an overt homosexual.

We talked as we ate. "Jeff, I want you to put your trust in Bradly-he's your friend, not your enemy. If necessary, he's been instructed to put his life on the line for you. Bradly is affiliated with the CIA, but his covert actions in The Genesis Protocol are separate, and hidden from these agencies. From now on, you will address your new friend simply as, well . . . Bradly. He's also instructed to assist you with any information or knowledge you need."

"Please, Scott, don't call me Jeff anymore."

"Then I'll address you as, Mr. Sandman?" he said.

I said nothing.

"Let me say a few more words to Bradly, and then you can continue your story of how horrible your life has been. From experimentation to reality, my friend here is about three decades into the future; physically, I mean. He's about 20 years ahead of his time. Only now, today, have genetic scientists considered the possibility of cloning a human, but with great skepticism, of course. Back in earlier days, as you may be aware Bradly, America wasn't necessarily everything people made her out to believe. The early days of innocence were a joke. Americans forgot who their government was, not the people, as it was written, but to the men who had money and power.

The voters and politicians hadn't the slightest idea of the true scope of this country's clandestine achievements. Unbeknown to them and our enemies, we had many secrets within, and some still remain-the Society is just one of them. This country was, and still is, the ultimate force on the face of this planet, and something to be reckoned with."

"So I might safely say that, The Genesis Protocol was the counterbalance of Adolf Hitler's little, brain-child dream of an Arian race?" Bradly asked.

"Bingo, Mr. Hues, but only after the war. Sandman was raised, taught and trained on an exclusive military base. Being cloned from the dead, they thought they could mold him into the ultimate soldier. Do you remember those days, Sandman?"

"I can't forget them," I simply said.

"I'm sure you can't. Okay, I've said what I wanted to say. Now, Mr. Sandman, I want you to tell us more-please, continue with your story for Bradly and me. Besides, I have only limited records of that period, and Bradly knows nothing-please tell us more."

They sat back and waited patently, while I swallowed a piece of my steak, and after a few moments of watchful silence, I started where I left off.

"I curse my growing years, and I'll never forget my most miserable days. I remember the woman that they made me call, *Nonny*; the bitch that helped raised me.

And I remember *Colonel Peck*, one of the men who taught and trained me. Nonny was a nasty, smelly old bitch, which, without my regard, used to masturbate as she watched the young soldiers march by through a windowpane looking over the kitchen sink. And Nonny was mean as hell. She used to slap my face a lot whether she was mad at me or not, and then call me a little, useless mutant devil. I must've been about six years old at the time; she was the first person I ever learned to hate. Even as an older boy, I dreamed of killing her.

Now, Colonel Peck; he was different, but I hated him too-mainly because he was such a callous motherfucker, but I didn't want to kill him. My education was restrained in the sciences, but he taught me about stealth, survival, planning and warfare, killing and firing weapons with quickness and precision.

He made me a ground soldier before I was even eighteen, and sometimes he told me he was proud of me-that's why I didn't want to kill him, especially after a year later, when he promoted me to First Lieutenant.

But it wasn't until I was twenty-one, that I knew there was something special about me. When I was twenty-one, *Dr. Simon Clark* sat me down and explained everything to me. He told me I was a clone. At first, I didn't believe him. I thought it was a weird joke until he showed me some files and pictures. It changed my life forever, because from that moment on,

I felt like some type of biological, subhuman being-someone who didn't belong in the world with ordinary human beings.

I felt like I was just some damned bio-machine. Dr. Simon Clark offered me a stay for four more years in the Marines. I stayed and continued to exercise everyday to build my strength. I succeeded in all the advanced military training and beyond.

Colonel Peck, as I said before, was instrumental in teaching me to fire a gun. I became better than him and was soon the best and fastest marksman on the base, no matter what type of weapon I used. I even set a new record in the marine's marksmanship. After I spent four more years in the service, I was offered a choice to go or stay. I chose to go, and at twenty-two, I was on my own. I received $300 a month and was given a mentor; commonly known as my caseworker. After two years of odd jobs-the last job being a real joke-I moved from a shelter into my apartment. Then the nightmares began.

My caseworker gave me instructions to see some doctors who were supposed to cure me. But now I realize that all this bullshit was a mistake from, The Genesis Protocol."

"You're right, Mr. Sandman, and my apologies for your nightmares. That was an oversight from the late director, Dr. Simon Clark; but I can assure you that there will be no more nightmares, bullshit or hidden secrets in the future."

"Then why am I here now? To tell me my nightmares are gone? If you brought me here to simply apologize, your time has been wasted. Now, if you'll please excuse me, I'll just be on my way," I said as I stood up and slid out my chair.

"*Sergeant Cooper*, would you come in here," Scott said in a superior and sturdy tone.

Within a second, Sergeant Cooper, who was a *very* big man, was standing in the threshold of the opened passage to the dining room. He stood at attention and waited for a command.

"Sergeant Cooper, please show this man to his seat."

Without hesitation, Sergeant Cooper started to walk aggressively in my direction, and likewise, without hesitation, I quickly sat back down in my seat.

"You're not going to let me go-are you?" I asked Scott.

"I'm afraid not. Not just yet, Sandman, I have important plans for you.

"Up to this time," Bradly interjected with a subject I had very little knowledge. "Mr. Sandman, I have been studying and supervising the mapping and sequencing of your personal genome. We have succeeded in compiling a complete, physical arrayed genomic library of all your cells. We also have encrypted your genetic code in our computer database."

"Scott, what the hell is this man talking about?"

"Sandman, what Bradly is trying to say, is biologically speaking, we know every cell in your body and how they work. We comprehend every position and function of every atom, and the base-paired molecules that make up every molecular unit of your DNA and RNA strands."

"Okay, Scott, now you're talking like Bradly-I didn't understand one word of what you said."

"Okay, Sandman. Let me see if I can make it clearer for you. We can *grow* a new heart, a new set of lungs, or a new liver, kidney, bones, and blood for you. You name it and we can make it."

"Okay . . . okay, I can buy that, but you guys cloned me in the first place, so what's the big deal now?"

"The big deal now is, in essence, we can mutate you," Scott said as a matter of fact.

"Mutate me! To hell you say! What do you think you're going to do, give me two fucking heads or something?" I shouted, and all this time, Bradly was just sitting back in his chair with a big smirk on his face that really pissed me off. "What the hell are you smiling about, Dr. Frankenstein! You think my life is just some kind of big fucking joke? Scott, you want me to call this bastard my friend? I've got a mind to just kick his ass and bounce his balls off the walls right now!

"That would be a bad mistake. This bastard, Sandman, is going to help change your life-for the good," Scott replied.

"And just how is Bradly going to do that?" I asked indignantly.

"Well, for starters, he corrected our uncertainty about those little, nasty nightmares you've been having. Remember when I asked you how you slept last night? How did you dream? You told me that you had a nice dream-quite pleasant from what I gathered. Tell Bradly thank you, Mr. Sandman, because it was he who cured your incessant nightmares."

"How did you do that, Bradly?" I asked, turning my head.

"It's called, human gene therapy. I found a defect in the genetic code from your brain's genome. I cultured some new brain tissue in vitro, and made the corrective replacement gene solution. Scott then injected the correct protein solution into the stem of your brain. Although I'll admit, I didn't think it would work as fast as it did.

"Hold it right there . . . when did you poke me, Scott?"

"Come on, Sandman; the last night I slept at your place."

"Then you cured me while I was drunk asleep?" I had to ask.

"Come the morning, you were none the wiser," Bradly concluded.

"Damn, Scott and Bradly, I'll have to say this, you guys are good."

I studied Scott and Bradly's faces when I looked down at the table and noticed that their plates were clean and I had hardly touched mine. Scott knew my plate was cold. "I see you haven't had time to enjoy your meal, would you like Henry to warm it for you?"

"Could you, please?" I asked politely.

Scott pushed that button again and Henry came out from behind the swinging doors.

"Henry, would you please refill our glasses and reheat Mr. Sandman's plate?"

"Yes, sir," he said. Henry refilled our glasses, took my plate and disappeared behind the doors.

"Buy the way, Sandman, did I tell you the entire fourth floor is yours? It has, of course, your private dining room and it's open to you all night and day. I have three chefs. You've already met Henry; tomorrow you'll meet *Carl* or *Dale*.

They each work out their own hours, so when you call for the chef, you'll never be quite sure which one you'll get. But I will assure you, all three make excellent meals."

We sat at the table for about a minute without saying a word, and then Scott finally broke the odd silence.

"Sandman, I what to change your name permanently, because in a month or two, Jeff Dunbar will no longer be."

"Ah yes, you're going to mutate me."

"Mutate is such and ugly word, Sandman; let's just say we're going to make you better than what you already are today."

"Scott, you've already cured me," but Scott just ignored my words.

"How does Christopher Lance sound to you?" he asked. I was dumfounded when I heard him say that name.

CHAPTER TEN

"How did you know that name? Did I ever mention that name to you before?"

"Sandman, to your impending disbelief, you have been watched by this Society, during and after you were set free. And, in light of your new girlfriend, I might suggest that you use your new name: Christopher Lance."

"Not quite, Scott. I want my new name to be: Christopher, Sandman, Lance."

"A very good choice for your middle name, it'll take on a new connotation-like the sand slipping through an hour-glass, a measure of time. I kind of like that.

I'll have your new identifications before you leave."

After Scott approved my new name, Henry came through the doors with my steak and I started to eat. "If you'll excuse us, Sandman, Bradly and I will visit in the guest room while you finish your meal. Please come and join us when you're through."

I scoured my plate like a ravenous pig and within minutes, my dish was clean. Bradly and Scott were sitting in the guest room and talking to someone new. As I walked in, I unknowingly said jokingly, "So, are you guys talking about me?"

"In this facility, Christopher, you are all we ever talk about," Scott replied. He saw me looking at the new person wondering whom she would turn out to be.

"Christopher, I'd like to introduce you to, *Dr. Karen Simson.* She specializes in human biology, genetics and anatomy." Dr. Simson stood up to shake my hand.

"Am I privy to your conservation's about me?" I asked while shaking Karen's hand.

"Of course you are, Christopher, and it's important that you share your thoughts and feelings about what's to become of you," Karen replied.

"Christopher, help yourself to the bar and make a drink," Scott said.

I made myself a double scotch with no ice and began to talk. "So, if you folks would be so kind, I would appreciate your candor and tell me want you're going to do with me. So would you please just be honest and enlighten me."

There was a pack of cigarettes and a lighter on the end table next to my chair. I lit a cigarette, took a drink and began to look at each of them, one by one, waiting for the first to speak.

"Sandman, Dr. Karen Simson will be a part of our team. She will be examining you very closely as we proceed. We plan to change you into a better man, unlike any man a person has ever seen," Scott said with a serious smile.

"Scott, I wished you'd cut the bullshit and just get to the chase."

"Okay, Sandman, but hold onto your chair. This Society is going to take you into thc 23rd century."

"You're going to do what!"

"I said we're going to extend your life into the 23rd century." Scott replied effortlessly.

"So this is your plan. This is what you're going to do to me. How?"

"We're going to change your genetics, which will in turn, slightly alter your anatomy. After extensive testing, and through successful experiments, our statistics predict that you'll live approximately another 200 or more years and that, my friend, will take you at least, into the year 2210," Bradly answered me.

"What do you mean by changing my genetics and anatomy?" I asked, rubbing my hands together nervously.

"Well, Sandman, let's go back to your nightmares," Scott replied. "They were caused by a malformation of genes in a specific part your brain. There is a subset of genes that control your emotions and psyche, which lies in a lobule near the base of your brain. Bradly discovered this anomaly while studying your genome, especially in the area that I just described. With a restriction-cutting enzyme, he was able to fragment that part of your brain's DNA. He found a group of neuropeptides, which was incorrectly organized and substituted it with the correct sequence of nucleotides-a new gene-which in turn, transformed your diseased brain cells into healthy cells. This transformation is why you had a nice dream. You're not understanding any of this; are you?" Scott concluded.

"Huh?" was my reply.

"Sandman, let me put it this way: Bradly used gene therapy to cure your nightmares."

"Okay, Scott-gene therapy. It corrected my bad dream cells to good dream cells, but what's that got to do with my anatomy?" I asked.

"All right, old buddy, I think it's starting to soak in. But now, we want to go a step farther. We want to modify particular parts of your brain and your body, which we know to be self-regulating. By doing this, your body will improve remarkably by continually making good cells to replace the bad cells. Did you understand what I just said? Scott asked me.

"I think so. And through this process, my anatomy and body will change?"

"Exactly," Scott said.

"But how will it change?" I asked Scott.

Scott turned to look at Karen for her explanation. "Christopher, let me see if I might be able to describe to you some of the physical changes you'll experience after your operation. First, your . . ."

"Operation!" I exclaimed with alarm.

"Not like you think, it wouldn't be anything like you envision," Karen said.

"Then please inform me, because I know, when doctors operate, they put you asleep and starting cutting away, just like I did with my steak."

It's not going to be that way, Christopher, because when we do operate, you'll be awake, at least for the first one," she replied.

"The first one! Okay, now I'm getting really worried, you're going to operate on me while I'm awake?"

"Just on your brain. Please, Christopher, sit down and relax; let me explain?"

"You go right ahead, Dr. Karen Simson, tell it to the human genie-pig," I said callously.

"First, and most importantly, we're going to start the operational injections in your brain, and for this reason, you must be awake. We may have to ask you some questions about the various sensations you're feeling. We do not want to damage your brain," Karen said.

"Oh, hell . . . fuck, sounds swell to me, I can hardly wait," I said with frustration and anger.

"Sandman shut up with your bullshit and try to listen without the stupid comments. Continue, Karen," Scott said calmly.

"Okay, Christopher, the first operation will start in your brain. We're going to open many pathways into your orbital regions. In particular, we'll start with your hypothalamus, and replace the cells with genetically altered stem cells. This part of your brain controls, along with other aspects, your basic rhythms of life. We are 99.9 percent certain that the new hormones released from these cells will re-regulate and increase the longevity of all the cells in your body. Your organs, muscles, nerves, tissues and everything else, will be uniquely transformed into what we call, super cells; like a super human being," Karen said with a twinge in her explanation.

"Will I be able to fly and have x-ray eyes?"

"Please, Christopher, this really is important, and the details are very complicated; but not impossible. If the procedure is successful, we will continue to give you injected gene therapy treatments. Let me tell you some of the things that will be altered in your anatomy."

"Okay, Karen, you have my full attention, this is beginning to sound serious, I promise, no smart-ass remarks."

Scott and Bradly were simply exchanging looks, and drinking their scotch while listening.

"Okay, Christopher. Your brain will become slightly larger, more convoluted and complex. As time goes on, you will gain more intelligence-to what degree, I can't say. Your skin will thicken, your heart will become stronger and slightly larger, your healthy cells will continue to divide for a very long time and your immune system will be able to fight infections, viruses, bacteria and many other diseases, without impunity. Your eyesight will also improve; you'll see as no other has seen. Your body will eliminate oxidative stresses and free radicals, which are contributory factors in aging. Your bones will become stronger, and your muscle tissues will give you the strength of ten men. I could go on and on, but I think you realize what all this implies. Do you?" Karen asked.

"Yes, I do. But hear this; if I become a side-show freak or something-I swear to God, I'll kill you all, including myself."

"Just the opposite, you'll be the ultimate male human species," she said.

"Will I still remember who I am and what I look like?—Or will I become someone completely new?"

"You'll retain all your memories and more," Karen said.

"You mentioned that my skin would become thicker. Does this mean knifes won't be able to cut me?"

"Oh, no. You will still be mortal. However, a simple cut, say, on your arm, might just leave a small incision. But to an ordinary man, the same cut would require medical attention; otherwise, he'd bleed to death. Although, your skin will be thicker, it'll still retain its soft and supple feel and appearance. And here's another point; your feeling of touch will be heighten. Here is another aspect that will interest you; if someone were to shoot you with a .22 caliber handgun, the bullet, in all probability, would just bounce off your chest. Of course, you would bruise."

"Damn . . . that's awesome. But, what did you say about my brain?"

"You will grow a thirst for knowledge and information," she simply said.

"Will I grow taller?"

"Slightly. Your height should increase by about one or two inches. And your face may look a little younger, with fewer wrinkles. Your overall mass will increase. Perhaps you may gain another 40 to 60 pounds. And as I said before, your strength will increase to about ten ordinary men. But I must make it absolutely clear; you will *not* be immortal."

"Okay, Karen, I understand. And Bradly said that I would live another 200 years?"

"200 or *more*," Bradly answered.

"Holy shit! But wait a minute-this means I'll out live all of you-what will I do then?"

"That will be for you to decide, Mr. Sandman," Scott answered.

"Could it be that I might become the new director of this Society?"

"Time, unfortunately, is the only tool as a test to measure failure or success," Scott replied.

"Okay. Are we through talking?" I asked, not wanting any more information.

"Yes, and it's getting late; it's ten o'clock and by tomorrow morning you'll need all the rest you can get, because later we'll get started. I'd also like to suggest that you give Darlene a call and tell her that you won't be seeing her for a month or two. Give her any explanation you wish, with the exception of where you are and what's happening. You're conversations will be monitored, so don't talk stupid, and for your information, you've just been sworn to secrecy," Scott concluded, and then added, "no one should know of this place; do you understand me soldier?"

"Yes sir."

"Sergeant Cooper, please show this gentleman back to his chambers."

Everybody got up and went in different directions, and I was escorted to my fourth floor apartment. Sergeant Cooper walked me up to the main entrance of the doublewide steel doors. After taking a few steps forward, I stopped and turned around just in time to hear the sound of dead bolts locking. It struck me hard with a little knot in my stomach; I was indeed a prisoner in this plush, and hopefully, temporary sanitarium. I went to my living room and poured myself a drink . . . watched a little TV, and thought about calling Darlene.

I sat in a chair drinking scotch, while watching the television and thinking what I'd say to her. Feeling it was time for a cigarette, I looked around and saw a fresh pack with a lighter on the end table. Gazing at a wall while I lit my cigarette, I saw one of those little red buttons and thought about pushing it just to see what would happen. Probably not a good idea, I thought. After watching the weather, I started feeling nervous but was uncertain why-should I try and sleep after I call Darlene? What time should I stop drinking? When will they wake me tomorrow; what will they do? If I was back at my apartment, I'd stay up all night drinking, but that didn't feel appropriate tonight.

I decided to switch to beer for my last few drinks, but discovered I couldn't find one, so I walked into the dining room and pushed the little button on the table, which would summon the chef. Within half a minute, and to my slight surprise, a chef walked in through those double swinging doors.

"Good evening, sir. My name is Carl, I'll be here until 5:30 a.m., how may I help you, Mr. Lance?"

I like the sound of that, I thought. "Could you get me a few bottles of beer and something to help me sleep tonight, Carl?"

"Yes, sir, any particular brand?"

"Of pills?"

"No sir-of beer?

"Oh . . . surprise me with something imported, and don't forget the sleeping pills."

"Right, sir, give me a few minutes and I'll be right back," Carl said.

I sat at the dining table and waited for Carl. After five minutes, he came through the double doors with a small cooler with three imported German beers inside, or so he told me. He handed me two pills, "Take these a half-hour before you go to bed," and then excused himself while departing through those doors again. I opened the cooler and grabbed a beer, sucking it down as I walked back into my living room. I sat back down in the chair facing the TV.

Setting next to me was a telephone on top of the end table and next to it the TV remote. I looked at my watch-it was almost midnight. As much as I hated too, I decided it was time to give Darlene a call.

"Hello?" she said in a very, sleepy voice.

"I'm sorry I woke you, Darlene, this is Christopher. I know it's late, but I had to call you."

"Is anything wrong?" she answered.

"No. But I wanted to call you to say that something important has come up.

"Something important?" she asked unsure.

"Yes, I'm going on a business trip."

"A business trip? How long will you be gone?"

"About a month, maybe two, but as soon as I get back I want to see you again, are you okay with that?"

"Yes, of course, but where are you going, Christopher?"

"I can't tell you that, Darlene, but I have an offer to write a book-one chance in a lifetime, but I'll be back soon."

"Are you leaving the country, or just to another state?"

"I'll be leaving the country, but I promise to try and give you a call while I'm away."

"Christopher . . . are you trying to dump me?"

"No. God, no, it's only business. I have a good chance to establish a contract to write a book. However, I can't disclose the nature. If I talk about it to anybody, it would be a breach in the contract. I hope you can understand."

"Christopher, I'm a writer and editor, I understand these types of things."

"Thank you for understanding, Darlene, I'll see you when I get back."

"Christopher?"

"Yeah?"

"I love you."

"I love you too. I know you have to work tomorrow, so please, go back to sleep. Dream about the next time we meet and the days and nights we'll have together-just you and me."

"Christopher . . . take care of yourself."

We said our good-byes and hung up the phone, and now all I could do was think about the future and what or where I was really headed to. I seemed to find myself in some type of philosophers' daydream. Ultimately, it transcended into thoughts about the universe, God and my life. I know I was created by human beings, but was it the doing of God who directed them to create me?

I remember back at my old apartment when I went into the bathroom to go pee. I saw a small centipede struggling to find his way free from the shower floor; and that little bug seemed very repulsive to me. After all, it was in *my* shower stall. I took a piece of toilet paper and squished him, threw him in the toilet, and then I started to pee. Looking down, I saw that the centipede was still alive and crawling on the toilet paper, trying so desperately to save itself; and all the while I was peeing on him, I knew it had no hope. When I was through, I flushed the toilet and watched the water swirl around and eventually, taking that little bug to its inept death. After the little creature was gone, I started to feel emotional. After all, it was alive, and it must of, had some sense of responsiveness, however little that might be. But was it God who directed me to squash that little bug and pee on him, while all the time he had no idea what or where he was really headed to? It made me start to think: was there really any difference between that little bug and me?

I stole another beer from the cooler and grabbed the remote to the TV. I turned the channels until I found the news. I watched for a little while until I became bored, and decided to look out a window but saw nothing but black. I tried a window on the other side, and there was one window that would allow me to see the small blinking lights of the big city far down below.

Looking at the city from such a distance made me kind of sad. Down there was my life, Darlene, Billy and even the bad boys of the night. The contrast from down there and up here was like the difference between a stormy night and a shinning day. Down there was destruction; up here was reconstruction; or at least I hoped.

It had been such a long day-my mind and body was exhausted and desired sleep. I finished my third and last beer and walked into the bedroom.

After undressing and laying my clothes on a chest at the foot of the bed, I crawled in, turned a lamp out and snuggled under the covers. Unlike my bed, this bed felt so comfortable that sleep came within minutes.

In the afternoon, about 1:00 p.m., I awoke remembering only bits and pieces of a very pleasant dream.

I stretched my arms and legs while yawning. The feeling was so peaceful that all I wanted to do was lie in bed and digest my waking dream. It was fragmented into different parts and my brain wanted to put the pieces together in some meaningful way. It had something to do with Darlene and me. So, in bed I laid for an hour or more reassembling my dream and taking comfort in the soft, white silky sheets. After I masturbated, I did my bathroom routine, and then found myself sitting at the dining room table, dressed in expensive clothes and pushing the button that would summon the god of nourishment. Through the double doors, came a man in white, who addressed me with a smile and a friendly voice.

"Good afternoon, Mr. Lance. My name is *Dale*. What can I get for you?"

"Dale . . . yes, could you bring me some coffee and breakfast?"

"Anything in particular, sir?"

"Surprise me."

Dale came back shortly with a carafe of coffee and condiments, and then left to fix my breakfast. I sat and sipped on my coffee and smoked a cigarette while waiting for my meal. Before I finished my second cigarette, Dale came through the doors and served me steak and eggs with whole-wheat toast and a tall glass of orange juice and milk. Of course, the meal was great and it was just what I needed. After eating, I pushed the button and Dale walked in and saw my empty plate.

"I trust you enjoyed your meal, sir."

"That was excellent, Dale. I never knew that eggs could taste that good. Thank you."

"My pleasure, sir."

"Listen, Dale, how can I talk to Scott? Do I have to push one of the red buttons on the wall?"

"You could if you want, but those are really just for emergencies. To talk to Scott, push the green button on the top of the telephone, sir."

Dale was cleaning the dining room while I was sitting in the living room waiting for a reply after pushing the green button.

"Good afternoon, Mr. Lance. This is Sergeant Cooper; how may I help you?" he said on the other end.

"I'd like to talk with Scott," I said in a polite, but assertive tone.

"I'm sorry, Mr. Lance, but Scott is out of the country away on business. I'll send Bradly up."

Then the phone went dead. After thirty minutes or so, Bradly and Karen walked in.

It was going on 4:00 p.m. and I'll have to say, without the freedom to come and go I was getting tense. But that went away quickly when Bradly told me that today we would get started.

"Is there anything I can do to help you?" Bradly said, looking at me, and then Dr. Karen Simson.

"I'm not sure, Bradly, but I'm feeling high-strung and apprehensive about today. Right now, I need something to help keep my mind in a state of sanity. It's not everyday that I experience the strange distress I'm feeling now."

"Okay, Mr. Sandman, I understand," Bradly said as he reached into his pocket. "Here-take these pills; they will help to minimize your unsettling tension and stress. Then I'd like for you to come with me. We have an appointment at six to start your treatments."

"Christopher?" Karen said. "Please let me assure you that tomorrow will go just fine. I am considered one of the best neurosurgeons in the entire world. You are free to ask me any questions that may be bothering you."

"You said that I would be awake through the whole surgery, but will I feel any pain?" I asked Karen.

"Absolutely not, Christopher. As a matter of fact, I plan on stimulating a part of your brain that will send you into a state of utopia for most of the procedure. However, occasionally I'll bring you back to your normal state so we might ask you a few questions."

"Okay, but what are my chances of surviving without you folks turning me into a cloned clown?"

"Your chances are very good that you will be just fine, if not better. I give the success of this operation a probability of 99.9 percent-and it just doesn't get any better than that," Karen replied with a strong surge of confidence.

"Would anyone care for a glass of scotch?" I asked.

"There will be no drinking today or tonight, Mr. Sandman. This procedure is too risky with even the slightest level of alcohol in your system tomorrow," Bradly said bluntly.

"You mean I can't even have a drink?"

"Not on your life. Now, please, just take the pills I gave you and then I want you to follow me," Bradly said. I did as Bradly asked while Karen watched. I swallowed the pills without water.

"I would've been more than happy to get you a glass of water," Karen said as she slipped a tongue depressor from the pocket of her white lab coat. "Now, open your mouth wide so I can look inside." I did as she asked, and then she probed my mouth with the stick. She looked under my tongue and all around until she was satisfied that I had indeed, swallowed all the pills. "Please don't misinterpret my actions, Christopher, I just wanted to make sure. Now, if you gentlemen will excuse me, I have to leave," she said as she walked out of the room.

"What were those pills you gave me, Bradly?"

"Two were to help relieve your stress and the other was to prepare your body for tomorrow."

"What do you mean, prepare my body?"

"One of those pills contained some special enzymes to help re-regulate some specific parts of your brain . . . but nothing you should worry about."

"Will I start feeling unusual or strange?"

"No, Sandman . . . just relax and follow me-it's time to go."

Bradly escorted me to the elevator on the inside of the steel doors and pushed a button. It lit a bright green.

"Have you tried to use the elevator yet?" Bradly asked me.

"No," I replied simply.

"Did you notice the bright green light?"

"Yeah."

"In the fraction of a second while I was pushing that button, it read my fingerprint and blood type, then compared them within the Society's database of classified clearances. It recognized that it was I, Dr. Bradly Hues, and thus sent the elevator. You, however, Mr. Sandman are not cleared for anything-yet."

Until I heard the word, *yet*, it felt as if Bradly was somehow, and in some way cautioning me, as if I were a dimwit. He might as well have been pointing his index finger at me, and waving it no-no. I imaged that his finger would be missing because I would have bitten it off. But, just hearing that final word, "yet", seemed to bring some kind of encouragement. Deep down inside, I knew that in time, I would acquire every available clearance known to this domicile.

When the elevator doors opened, I smiled at Bradly and waved him in first. We both turned around to look at the buttons on the inside when I saw Bradly push a button labeled: negative two. Again, a bright green light scanned his finger.

Bradly turned to look at me with a smile and raised eyebrows. I couldn't quite understand what that look was for. When the elevator door opened again, all I saw was a white corridor. Bradly led the way.

"Where are we going, Bradly?"

"Please, Sandman, just follow me."

As we walked through the corridor, I noticed there were cameras following our every step, and watching our every move. Along the way I saw many different colored doors labeled with symbols that meant nothing to me. Finally we stopped and turned to a pair of blue doors.

Bradly pushed another one of those buttons and the doors disappeared immediately into some recesses built inside the walls. After passing through the doorway, the doors closed behind us as quickly as they opened. The smell of cleanness was the first thing that overpowered my senses. There was a deep smell of antiseptics, ammonia and other cleansers that assaulted my nose and caused me to sneeze. The room appeared to be fashioned much like a hospital, with people dressed in pure white gowns and going about their business, whatever that might be. The walls of the room were lined with equipment and instruments that I imagined monitored a body's vital signs, with clear plastic tubing connected to needles and endless other things. The sounds of beeps, bubbles and air compressing were almost louder than the worker's feet shuffling around. I turned to look at Bradly and all he did was simply look back at me.

"Well?" I asked.

"Well what?" Bradly replied.

"Are you going to tell me why the hell I'm here and what they're going to do to me?" I questioned him in a loud and wavering voice, which drew no attention from the people working around.

"Relax," he said, "all they're going to do is prep you for tomorrow."

"Prep me how?"

"You'll see."

Bradly escorted me to a smaller room and introduced me to a lady he called, *Nancy*.

"Good afternoon, Nancy. I'd like to introduce you to a very special man whom you've been waiting to meet. Nancy, this is the infamous, Mr. Sandman you've been hearing about."

"Very nice to meet you," she said holding her hand out.

As we shook hands, I felt my face grow warm. I found myself feeling kind of intimidated by her attractive looks and apparent intelligence. Her eyes were deep blue and her smile was bright and wide. Although she was wearing a cap and gown, I still noticed she had red hair and a pretty body to match. Her breasts filled the top of her gown and her legs were slender, long and sleek. Her face was smooth and clean as if some very talented artist sculptured her. She appeared to be in her early thirties, although it was hard to tell. The bottom line was: she was a knockout who would weaken the knees of even the worldliest of men.

"May I call you Christopher?" she asked.

"Call me bozo, goofy, dumb-dumb, or even stupid if you like," I replied infatuated, which I suspected she knew.

"I'll stick with Christopher, if you don't mind."

"You're in good hands," Bradly interrupted, "I'll see you later, Sandman; have fun."

Bradly turned around and walked away, and without taking my eyes off Nancy, I just said, "Okay."

"All right, Christopher, are you ready to get started?" she said, breaking my trance from her big, blue beautiful eyes.

"Let's do it . . . where do we start?"

CHAPTER ELEVEN

"Okay, Christopher, first; I want you to take a shower. Over there in the corner is an enclosed shower. However, this shower stall was designed to remove all your body hair and the first few layers of your skin."

"Are you telling me to take a shower in acid or something?"

"Believe it or not, Christopher, but you're close to the truth. The liquid that we'll spray on you is an organic acidic solution. The purpose of the shower is to sanitize your body and remove any bacteria on the outer layers of your skin. Are you aware of how many microorganisms, germs and bacteria reside just within and on your hair and skin?

Most all of them are harmless outside the body, but inside your brain, they would wreak havoc. This is part of the prep for tomorrow's operation. Now, before we start, I'll need to insert some special lens' caps into your eye sockets for protection. Please, lie down on this table-bed while I insert the lenses."

"God, I just know this is going to hurt," I whined with already watery eyes."

"Just relax and lay back while I spread your eyes open with these clamps. But let me explain one thing to you before I begin. You're going to feel some pressure and hear some noise as I insert these things-don't confuse these sensations with pain," she reassured me.

Those clamps forced the top and bottom of my eyelids to spread wide open. It was a very strange sensation. Nancy put some eye-drops in and the watery feeling went away. Lying down on my back, I turned my head to see what she was doing. I saw her picking up one of those lenses with a tool that looked like a combination of a suction cup mounted on a round, small cookie cutter. She started with my left eye and I indeed felt and heard some unnatural sounds grinding and snapping somewhere in my head and between my ears. After she finished inserting those bubble-lenses in my eyes, and after removing the clamps, Nancy began to speak again.

"Now, that wasn't so bad; was it?"

"I've enjoyed myself in better ways," I replied.

"Are you feeling any pain?"

"No, but I feel like a walleyed fish."

I heard Nancy's slight chuckle as she said, "We're not quite finished yet."

"Oh damn. God help me, I wished I knew where we were going with this?"

"Christopher, in the most simplest of terms, I'm going to have to insert some cavity plugs for added protection. I'd let you do it yourself, but this procedure must be professionally performed-besides, you wouldn't be able to stomach it. Your ears, nasal, throat, anus and penis must be plugged," she said with a bashful, but honest laugh.

"You're shittin' me?"

"No, but that's next. After collecting a sample of your stool, we'll also need other samples: urine, blood, sperm and spinal fluid. We'll gather those samples after your shower, but for now, I must ask you to step into this cubicle for your enema and stool sample."

With a little bit of embarrassment, I followed her instructions.

After my enema, (which I'll admit, felt very refreshing and cleansing), Nancy proceeded to insert little earplugs and nose-plugs, including the anal and urinary plugs. As she was starting to insert the penis-plug, I grew an erection. Nancy continued to work and simply said, "Quite normal for a man your age . . . I think I'll consider that as a compliment."

"I'm sorry, I couldn't help it."

"That's okay, Christopher-you're going to need it later."

Nancy plugged-up my shit eating grin with a breathing type apparatus she placed between my lips. She instructed me to continue breathing normally. I was then given a washcloth and told to simply pretend to take a regular shower. Again, I did what she said.

When the shower started to spray, I felt a slight tingle all over my body. Before I even began using the washcloth, I started to notice little goo-like lumps falling from the top of my head. It was my hair. I felt the top of my head and it felt as smooth as wet ice, but without the cold. As

I continued to wash, I saw the hair on my arms; chest and legs just melt away. The curiosity got the better of me and I had to look. My cock looked younger to me, now being bald and all. But, at first and false glance, what seemed to frighten me worst was how small and short it looked. Only after spending some time investigating, did I realized that my penis was indeed, proportionally correct.

Suddenly, a blast of warm air swirled around me, drying my body, and for some unknown reason, I instinctively pulled the plugs from my nose and impulsively smelt my skin. I was taken by surprise by how my body smelt-fresh like a newborn, two-week old baby. I enjoyed the innocent fragrance.

I stepped out of the shower stall and pulled those bubble-ball lenses out of my eyeball sockets along with all of the other plugs still stuck in my body's cavities. I looked at my baldhead and face up close in a mirror. To my surprise, I had no eyebrows, and after Nancy wiped my eyelids with something, even my eyelashes were gone. After looking at myself in the mirror, I felt somehow, and in a strange way, that I was not created from this earth-and yet, here I am. And from what I understand, I'll be the quintessential step in the development of humankind; a more highly evolved, mutated kind of man.

"Does your reflection frighten you?" Nancy asked in a sincere tone.

"I wouldn't call it fright; I'd call it, not right. My reflection doesn't look right to me."

"Don't let your appearance deceive you right now. All your facial and body hair will grow back, Christopher. Now for your climax of the day-and I mean that in more then one way-would you please put this gown on and take this cup? Inside the room over there, is where you will deposit your sperm, hopefully into the cup.

There are pornographic movies to watch. When you're through, push that red button on the wall. Take your time, and enjoy yourself-this is as good as it gets."

I proceeded to the room and shut the door behind me. After a pleasant period of time, I came out with the cup, placed it on the counter and pushed the red button. It wasn't long before Nancy walked in and positioned the cup in a container.

"My goodness, Christopher, you must've had a very good time. I don't think I've ever seen a cup this full."

"Well, I did my best, but I didn't know how big of a sample you wanted." She looked at me and smiled for a second or two.

"Okay, Christopher, just three more procedures and then we'll call it quits. First, I want you to lie back down on the table-bed, because I need to draw some blood and spinal fluid. Second and third, we're going to scan your head with lasers so our computers can take precise 3D pictures of your skull; it only takes about twenty seconds. After that we're going to image your brain, using a very high resolution MRI, and then superimpose the two 3D images together into a true 3D model-right down to the nanometer. We'll need this information to download into the neurological O.R. robots that perform most of the precision work. Tomorrow, they'll put you in the head-cage and reference your skull and brain with the images we take today."

"I hope the hell all you people know what you're doing," was the only thing I could say.

"Trust me, Christopher, we do; and sometimes, all too well. Please lay back down so I can draw some blood," and after Nancy drew three vials, she continued, "Now, roll over on your side and curl up into a ball."

I felt her touching my lower spine and then it went cold. At first, I thought she was pinching me hard until I realized that it was the needle being slowly inserted to draw the spinal fluid. The whole ordeal took about fifteen minutes.

"I want you to roll on your back again and just relax awhile," she said after the procedure.

I did what she said. I discovered that the time afforded me to think about what I just did, and again, where I was headed to. What also struck me was the apathy in which they used me-to play on my ignorance. Never in my mind did I realize that this day was going to end this way. It was at this moment that I truly became aware that this was indeed, reality and that all the future events were going to be for real too. My mutation was soon to begin and only now did I recognize that this was not a game. My life is just as important to me, as life is to others. I didn't want to die anymore than anyone else.

But by divine destiny, or the tools of technology, here I am, waiting to know what is to become of me. But as Scott had said: "Time is the only tool as a test . . ." and now I believed in what he said.

"Hello, Christopher, I'm back. I'd now like to get started with the laser scans, as I said earlier, this is so our computers can use those 3D pictures to fashion a virtual image of your skull."

She moved my table-bed under a machine that covered my head. I was told to simply relax and keep my eyes shut until I heard her say, "okay". After about twenty seconds, I was through. That simple. Nancy then wheeled my table inside a room containing one of those huge steel donut-shaped devices-an MRI machine. Again, the procedure wasn't long and before I knew it, I was through.

"Time to get up, Christopher, your prep is complete. I want you to go back in the changing room and slip this body suite on," Nancy said with a professional voice and expression.

I followed her instructions and came back out feeling very different, again standing at the same place that I came in. Nancy no longer appeared to me as she did when I first walked in, and I think she felt the same. But I knew the real reason was because of me, a hairless-curiosity wearing this white body suit.

Just then, Mr. Bradly Hues strolled in after the outside doors quickly slid open, and then closed again.

"Well, Mr. Sandman, I hope that didn't spoil the rest of your day, but we needed to do this in a gradual way. In short, we needed your body fluids by catching you with your guard down, if I may use that cliché. The fluids we received from you today will be used to calibrate our machines and computers so that we have a starting point, (a referenced time-period in your genome if you will.) From this day forward, we'll monitor your progression. Are you ready to go?"

I looked back at Nancy and said "good-bye".

"May God be with you, Christopher, and I hope you have a wonderful life," she said, and then turned and walked away.

"What time is it?" I asked Bradly as we walked back down the corridor.

"It's ten o'clock."

"Why did we start this so late?"

"Well, today we chose to work from your time schedule. But I will assure you, all of that will change," he concluded as we stopped in front of the elevator.

"This time Sandman, *you* push the green button." I did and the elevator doors opened.

"I'm impressed," I said in a non-impressive tone and stepped into the elevator with Bradly. I pushed the forth-floor button and Bradly began to speak. "You're in the data base now, and we've given you some designated clearances."

"Cool."

"Of course, with time, your clearances will become less restrictive," Bradly said as we stepped through the elevator door again. And when we walked into my living room, Karen was there to greet me.

"Good evening, Christopher. I wanted to talk to you before tomorrow."

I sat down without saying a word while staring at her face to listen.

"Tomorrow, we start the procedure. I want to give you some idea of what we're going to do from start to finish, and I'll try to put it in simple terms.

First: we're going to genetically mutate you, as I said before, starting with your brain. I'll inject rectified, disease-free chromosomes into specific areas of your cranial lobes.

Strictly speaking, we're going to start at the very core of you brain, your animal brain if you will-the hypothalamus-where the basic rhythms of life evolved . . . and from this area, we'll work our way outward. Obviously, I can't stress enough about how critical it is when operating inside the brain. It is our Achilles heel, and hence, there can be absolutely no room for error. None.

The brain images that we collected today will achieve this goal. Second: we'll do basically the same with your body. Through gene-corrective therapy using nano-needle surgical treatments, we're going to insert new chromosomes in all your organs and throughout your entire anatomy. However, for you and myself, this procedure is not as critical as with your brain, and so I believe it will be less unnerving. Do you have any questions?" Karen asked.

"I'm all out of questions. The way I see it is this: I'm a captured clone and through this new miracle of medicine-which everyone has convinced me of-I will be transformed, or mutated into a new hybrid of a human being. Will I still belong to the Homo sapiens species?—Or will you have a new name for me when this is all over?"

"Why yes, strange that you should asked. Would you like to hear it?"

"Sure."

"Homo genesis sapiens," she said.

"Huh, kind of makes me sound important."

"You will be; probably more than you'll ever know," she replied in a collected tone. "Okay, Christopher, let's end this discussion-we have a lot to do tomorrow. I want you to take these two pills an hour before you go to bed."

"What are they?" I asked, as if I really needed to know.

"They'll help you get a good nights' sleep. The operation begins at nine in the morning. Please, don't stay up any later then twelve tonight."

I completely forgot that Bradly was behind me and to the left, still standing. He followed Karen and on his way out, without turning his head, I heard him say, "good-night."

They allowed me to have my cigarettes, but booze was out of the question. I spent about an hour or so watching that big TV until about 11:30 pm, and then took my pills and went off to bed. After a few minutes, and a fleeting thought of Darlene, I had left reality and into my realm of new and fascinating dreams.

"Wake-up, Christopher, its seven o'clock in the morning. This is your big day," Karen's voice echoed and bounced around my head. I sat up in bed and waited just long enough for a morning erection to subside. I was surprised and a little dumbfounded to see her actually standing in my bedroom with clothes in her arm.

"May I put my robe on?" I asked while rubbing my drowsy eyes.

"Put this on instead," Karen interjected, handing me a gown and a pair of silly slippers.

"Where are we going?" I asked as I clumsily put on the garb.

"Today's the day, Christopher. We're starting your initial operation. We'll make a short pit stop on sub-level two for a little spit and shine. Then we'll go to sub-level three for you're . . ."

"I know, I know; for my operation."

"Right. Okay, Christopher, please follow me."

As we approached the elevator, I skirted around Karen because I wanted to push the button. "Please, allow me," I said, and was somewhat surprised when the doors slid open. We walked in and I then pushed the button labeled: sub-level two, but nothing happened.

"Nice try, Christopher, but you haven't been given authorization yet," Karen said, and then pushed the button.

When the elevator door opened again, the familiar sight of the corridor and eventually the blue door brought back my thoughts of Nancy.

But today, there was no Nancy. Instead, there was a small, frail, feminine-looking, aged man, wearing latex gloves, standing silently before me and waiting for Karen to speak.

"Good-morning, *Dr. Wayne Koloski*, this is Mr. Christopher Lance. Christopher, I'd like to introduce you to Dr. Koloski," she said.

I was pleased when he didn't offer me a latex handshake, but instead, just gave me a wink and a little smile. After seeing that, I thought, I would've rather had a handshake.

"Wayne, could you see to it that Christopher has his morning constitution, teeth cleaned and body showered?" Karen said.

"Of course, Karen," Wayne replied, "I'll have him ready in about thirty minutes."

I thought to myself; it's kind of interesting that in such a short period of time, how quickly one can start to see the dynamics of an organization and the hierarchy that becomes so apparent-especially to an outside observer such as myself. I had to wonder if they were aware of the fact that I was, in my own way, taking mental notes and pictures and spying back at them. For me, I thought, it was in my best interest and possible self-survival that I did this personal surveillance.

"If you'll follow me, Mr. Sandman, I'll show you the way to the bathroom for your pre-op cleaning," Wayne said.

"Oh, hell no! Not again. Not the enema and the bubbled eyeballs and those damn cavity plugs-no way-especially with *Dr. Wayne Latex*! This is cruel and unusual treatment, Karen; do I have to do that all over . . ."

"No?" Karen said just loud enough to interrupt my words to an immediate stop. "Just your typical bathroom routine and a simple shower," she repeated.

"What?" I asked.

"We'd like for you to brush your teeth and if possible, have a bowel movement followed by a normal shower before we proceed to the O.R., Christopher. Do you think you can do that?"

"Cavity plugs?" I asked frankly.

"None, just what I asked and nothing more." Karen said, as I caught a fast glance of her eyes rolling-as if I were just fucking around with her-which I wasn't. I was not going to let Dr. Latex shove plugs up my ass and inside the end of my cock-plain and simple. That part I would have to insist on doing myself.

While walking to the bathroom, I glanced over at the porno-booth. No lost opportunities in that little private cubical. However, as I continued to walk to the bathroom my thoughts started to become more serious and somber. This was really happening. It suddenly hit home to me how real it all was, and how truly serious these people are. For the first time in my being, I really comprehended the fear of losing my life. Through all of my little courageous adventures, never had I met an adversary, or advocate, like this. And I also must say, that putting my life in other people's hands is very hard for me to do. But starting today and the days that follow, that's what I'll have to do.

After my shower, that warm air came on and dried my body. When I stepped out of the bathroom, Dr. Latex Koloski instructed me to lie down on my back on a special O.R. table-bed with colored bags of liquid hanging from posts. After I did, he covered me with a sheet, poked some needles in me, plugged in some tubes, and then proceeded to wheel me to what I suppose, would be an elevator. While in the elevator, I happened to notice Dr. Koloski pushing the sub-level three button with his latex gloves on. He noticed me looking.

"The scanner can see through gloves," Dr. Latex said, and then added, "Marvelous technology-isn't it?" Not that I didn't think he was a smart man, I just read him as gay, and I never associate with people who enjoyed the same sex-I don't fit into their way of life. Suddenly it occurred to me, I was wasting time thinking about bullshit. My mind was babbling as the elevator door opened.

"Well, here he is Dr. Simson, I'll place him in position," Wayne said as he turned to move me around.

"Thank you, Wayne," Karen answered him as he wheeled me under some soft lights. After Dr. Latex had left, it took less then two seconds before many eyes encircled me-very serious eyes. They darted back and forth and up and down like little mechanical eyeballs. With surgical masks covering them, everyone looked the same, with the exception of one pair of eyes. It was her eyes, not her voice, which would talk to me. "Move him into the body cavity and secure him."

I felt like an astronaut lying in a horizontal form fitted bed. "How did you know how to make this thing?—I didn't lay in any molds."

"When you initially, and subsequently walked through the opened elevator door, you were scanned, giving us a time motion video, which we were able to create a three dimensional form. This form allowed us to build your cavity," Dr. Bradly Hues explained.

"Okay, Christopher, You're locked into your body cavity. As you probably may have tried; you can't move-and try not too for your own benefit," I saw Karen's eyes talk to me, "Now we're swabbing your head with a topical, clear gel which will anesthetized the layers of your skin-right down to your bony skull. It only takes seconds for the gel to osmotically dry and take effect, but the numbness will last for hours.

You're doing very well, Christopher. Don't get excited about what we're going to do now-it's the most important step in the procedure. We're fastening the head-cage to your body cavity-don't let it intimidate you."

When the head-cage was being connected, I was actually quite surprised to see that the device wasn't as medieval as its connotations implied. From my point of view, it looked lightly constructed, and shaped like a spider connected to a rotating shaft at the head of the bed. There were at least a hundred little tubular 'legs' with tiny cups on the ends, and with many wires.

"This is the most critical robotic device we have. It holds your head stationary within the cage, while it performs the necessary laser drilling and needle inserting.

The cage's mechanical, thin appendages guide the laser and the needle's initial insertion, and to do this, it must have many degrees of freedom, or to put it more succinctly, it must be able to move in any direction. We use numerous micro-servomotors which are controlled by either humans, or in most cases, by our compound computer, capable of light speed computation and transmission," Karen said.

"That's nice-very good, but why are you telling me this? Should I be taking notes?"

"I might also add, that the computerized head-cage is also capable of displaying virtual motion pictures of your skull and brain superimposed together in real-time.

Our image monitors display an ultra-fine resolution, much better then the best high-definition displays on the world market today," Karen replied.

"Cool," was the only thing I could think to say.

"Dr. Hues, would you begin inserting the skull pins," Karen asked.

"Skull pins?" I questioned nervously.

"Relax, Christopher, you won't feel any pain," she said reassuring me.

When the head-cage was fastened and confirmation given, I felt a numb, soft pressure positioning my head. Then I heard a noise kick on-it sounded like something spinning very fast, and then I smelt the odor of burning flesh, and finally watched the smoke wafting away in the air. Within seconds, I had fifteen simultaneous little holes for the skull pins to be nestled in.

I could feel little points of pressure when the robot started to insert the pins, and as it did, I counted them. He continued to work on me for a few more minutes, and then announced that he was done.

"He's all yours now, Dr. Simson."

"How are we doing, Christopher? Are you okay?"

"You tell me. I feel okay-am I okay?"

"You're fine, Christopher. Now I'm just doing a little calibrating and synchronizing and . . . hold it . . . and, there. There we are-up and running."

Just after Karen finished her sentence, I saw a thin gray tube snake out of one of the ends of an appendage and stare at me. I was temporarily confused. It took me a few moments before I realized it was mechanical. It moved as if it were alive.

"Don't look at the *twistube*, Christopher, look at the monitor over here and tell me what you see?"

I looked at the monitor, "I see my forehead; big deal. Oh . . . okay, now I see my ear. Wow . . . now I see my face-it looks like my reflection in a shiny bowl."

"Good. Do you see your face clearly?"

"Yeah, just like looking in a shaving mirror; but better. I can even see the little holes in my face where my beard grew."

"Excellent, Christopher. This is basically how the device works and how we see it. The legs, or appendages of the *spider*, as we like to call it, is connected inside the head-cage and can position itself within a nanometer anywhere on your head.

At that point, a much smaller twistube emerges which can see inside your head; and that includes your brain. It works like a global positioning system and our computer graphics has incredible zoom and resolution capabilities as well. Let me demonstrate very quickly. Do you see that tiny little mole next to the corner of your left eyebrow?"

"Yeah."

"Okay . . . now watch," she said as the twistube zoomed toward my face-the tiny mole changed into a huge looking, dried raisin.

"Looks ugly," I said, not sure what else to say.

"Please keep watching the monitor," Karen insisted.

To my horror, disbelief and fascination, I watched as my face started to slowly melt away. It was quite unsettling to watch it decay-like a fast motion movie of slow, decomposing flesh. The images kept devouring away my face until I could see my brain. After a few seconds, the monitor quickly flicked back to my normal face. When the pictures started moving toward the back of my head, Karen moved the monitor away from my sight.

"*Dr. Ringrose*, would you please secure the clamps?" Bradly said as the operating room lights intensified without hurting my eyes.

"Clamps! What clamps? You didn't say anything about clamps," I said loudly.

"Don't let it concern you, Mr. Lance, we have to secure your cavity to the table. It's for your safety and ours. Uncontrolled, or involuntary rotation could very well jeopardize not only this operation, but also your life," Dr. Ringrose assured me with his talking eyes.

"Okay, nurses and doctors, let's connect the patient-I need some vitals," Karen commanded in a confident and esteemed voice.

"Blood pressure at 105 over 70," a pair of eyes said.

"Heart rate at 80 bpm; respiration at 18," another answered.

"Pulse Ox at 99 percent."

"Zooming at 10 square centimeters," Karen said.

"No anomalies in brain waves: alpha, beta, gamma and delta."

"Receiving 99.4 percent node impulse reflections from the twistube's matrix-well within the limits, Dr. Karen," someone added."

She played with some dials and flipped some switches then waited patiently for a moment. "Alright everyone, his vitals look good, I'm ready to go in," I saw her eyes say as she sat at the monitor and the controls of the spider. Then, she spoke again. "Bradly, what's the command?"

"Computer confirms . . . Status: *genesis*," Bradly announced.

After I heard the word: genesis, they all started to celebrate quietly. I heard cheerful sounds of success and reassured confidence resonant. But the little celebration didn't afford me any confidence. All it told me was this shit really indeed, has never been done before. My first initial thought was . . . Homo genesis sapiens? When this is all over, I'll just be happy if I don't start walking on all fucking fours.

"Okay, folks. That's a green for go. Bradly, do you have coordinate conformation for needle insertions?"

"Conformation is complete, zero position needle reads delta x at 0.0000, delta y at- 6.2745, and delta z: -5.3448. Coordinates normalized . . . set and locked; you may start the twistube whenever you're ready, Dr. Simson."

"Okay, Christopher, we're starting the initial and critical phase for the needle insertions. The sensation you're feeling now, is the pressure from the twistube that's drilling a hole in the back of your head. When we come into contact with the outer most cortex, the micro-twistube, which is about the size of a needle will continue farther into the deepest part of your brain-your hypothalamus," Karen told me.

"It looked like a small, gray snake-how in the hell do you make it move like it's alive? Or wouldn't I understand?" I asked.

"I don't think so, Christopher. The twistubes are special because of their abilities-both of them-the one you saw earlier and the one we're using now.

They were designed using a low-density polymer with the physical properties of turning; bending and twisting effortlessly like a worm. The sides of the twistube walls are wrapped in an ultra-thin electrical and diagonal mesh matrix with millions of intersectional nodes. When the computer is instructed to make the twistube turn, it sends an electrical impulse to that area of intersecting nodes.

When the current makes its connection, the electromagnetic effects from the nearby nodes create a kink in the twistube's wall. It was discovered that this procedure, using computers, mimics the way a living worm's muscles contract, stretch, bend and so on. Therefore, kinking the twistube walls allow us to serpentine; making it flex, bend, turn, rotate and twist in any direction we wish. Just like a worm. What did you think we were going to do, Christopher, stick a straight, long needle through your head?" she said playfully with squinting eyes.

"Well, yes . . . I kind of thought . . . hell, I didn't know. You guys are the pros, I didn't even think about it before now. Karen, you've told me more then once what was going to happen to me. But please . . . from now on, spare me the details?"

"How's his brain activity?" Karen asked someone.

"Six percent increased in alpha frequency with small anomalies in gamma-no change in beta or delta," one of the nurses reported.

"Christopher, are you feeling any strange or unusual sensations?" Karen asked.

"No. Why are you asking?"

"Just try and relax as best you can. We're now threading the micro-twistube through the cavities and channels between the various lobes in your brain."

After a few more minutes of this, I started to shake involuntary with rapid jerks. I couldn't help myself. I couldn't even talk. But I was indeed aware of what was happening to me, including the noises and voices.

"We're in the hypothalamus," I heard Karen say impatiently, "We don't have a lot of time. Someone give me some vitals."

"Blood pressure at 80 over 40."

"Heart rate at 65 bpm."

"Pulse Ox is 86 percent," a voice answered.

"Good. Start injecting the surrogate genes," Karen told Bradly.

After a few seconds, Bradly replied. "Injection complete."

"Okay, let's retract the needle," Karen instructed.

"Clear."

Slowly, my involuntary convolutions started to ease away. The doctors didn't appear to be concerned and in a couple of minutes, I felt just as normal as when we began. However, I was still slightly troubled by the events that played out in my head. Karen started reading my vitals out loud and seemed pleased. "Vitals check. Do we have computer confirmation yet, Bradly?"

"Computer confirms . . . status: *Bravo*," I heard Bradly say.

"Very good, people; we're over the hump," Karen rejoiced.

"Is this shit going to happen every time you stick that goddamn needle in me?" I asked angrily.

"No, Christopher that was a one-shot-deal. You've passed the most complicated procedure of your first operation. By the way, I'm happy to see you're still alive. You could have died. Congratulations!" Karen said with a smile.

"Congratulations. Okay, I could have died. I feel honored. Thank you all, yes, thank you for not killing me. Are we done now, Karen?" I pleaded.

"You gotta be someplace?" she questioned my question softly.

"There's more, isn't there?" I had to say.

"Yes, Christopher, a lot more, but the rest of the injections will be performed by the computer and the robotic head-cage. No human intervention. The remaining brain injections will take approximately six hours, but we'll stick around to cheer you on."

Through out the whole six hours, I laid strapped to the table wearing my immovable headgear. Without even the smallest break, it went on, one after another.

First the laser, then an injection, again the lasers, then more injections, and on it went, over and over again. But it didn't bore me in the least. As a matter of fact, it scared the hell out of me.

All through this procedure, I was constantly perplexed by the effects I was experiencing: smells, sights, sounds and taste. I heard the sound of a snowflake falling like a comet crashing to earth. I smelled good food cooking, and then smelled spoiled eggs. I swear I even smelled the odor of black. I was gifted the sight to see the blinding lights of a billion stars all clumped into one. And I saw the darkest-dark, in the abyss of a Godless universe. Dr. Simon Clark's apple pie sure tasted good, but I didn't like Colonel Peck's boiled snakeheads. I felt the pleasure of ejaculation and the pain of losing love. On and on my experiences went. But something in my mind tried to tell me that my sensations were not true. Because of those needles worming through my brain, I became hypnotized and paralyzed like a zombie-there was nothing I could do. Bliss unified with horror. I didn't enjoy it in the least. Suddenly, my surreal senses were jolted back to reality. The operation was over. After everything was removed and disconnected, one of the nurses put a bandage-cap on my head.

"How are you feeling, Christopher?" Karen asked.

"I'm not sure . . . I mean, I don't know. There's something I want to ask you . . . I think," I responded, unsure of what to ask.

"Are you feeling pain?" she asked with concern.

"No."

"If there's anything else, please tell me?" she said.

"The sensations; my thoughts, sights and feelings. I experienced something that felt very unworldly through a series of dreams. Did you make me sick again? I don't want any more of those damn nightmares; please, no more bad dreams."

"Relax. Relax," she said while embracing my arms in her hands, "There'll be no more nightmares-those days have past.

What you experienced, Christopher was nothing more than hallucinatory manifestations due to the stimuli of cranial gray matter. It's very common and nothing to worry about. Did it feel like you were on the mother of all acid trips?" she asked.

"Is that what acid trips feels like? Now I know why acid freaks never know where they are. And they pay money for these hallucinatory trips? Assholes."

"Christopher, as soon as the topical anesthetic starts to wear off, you're going to start feeling some pain. Don't freak out and please don't scratch your head. We have medication that will take care of that. Right now, I want you to sit in this wheel chair—we're transferring you to the infirmary. You'll be monitored and watched throughout the night to make sure you're vitals remain okay. Then, after we determine that everything is indeed all right, we'll start on your body injections after tomorrow," Karen told me.

Karen wheeled me to the elevator and pushed sub-level one. I was taken to a private room and wired back to some vital monitoring systems. I laid in bed watching television while waiting for the pain. And the pain came. It started with intense itching and progressed to an intolerable pricking. I felt like a pincushion. About as soon as that began, a headache proceeded. It started slowly until it built up to an explosion inside, and then worked its way outward until my eyes felt bulged. The throbbing and pain was becoming more than I could bear and so I sought for help-I screamed.

"Help me! Nurse! I'm dying in here. My head is going crazy!" Within two seconds-about two minutes for me-a nurse walked in. "Help me, please!" I pleaded, "I can't take another minute of this pain!"

"I'm very sorry, Christopher, your help button was not at your side when the anesthesia started to go away," the nurse commented, and then approached me with a hypodermic needle.

"Is that the stuff Karen told me of; the medication that takes away the pain?" I asked begging like an unfixed-junkie.

"This shot will take all your pain away. Give it about five minutes and you'll be feeling just delightful," the nurse reassured me.

After she gave me the shot, it wasn't long before the pain did indeed go away. As a matter of fact, I remember singing some stupid songs and making up silly words as I went along. After about half an hour of silliness, I was dead asleep.

CHAPTER TWELVE

"Wake up, Christopher, I have some medication you need to take."

I opened my eyes and saw a nurse standing beside me. "I think I made an ass of myself last night," I said, and then took the pills she gave me.

"You sang some songs, but don't be ashamed-it was the medication. How's your head feeling?"

"Okay I guess-I don't feel any pain."

"Well, that's a very good sign. Your vitals are good, and your brain activity looks normal. Can you sit up for me-I need to take your head dressing off."

I sat up in bed and the nurse began to take the bandages off. "Can I have a look? I want to see the damages," I inquired.

"Sure," the nurse said as she handed me a mirror.

To my surprise, all I saw were small red dots, and not the hundreds of little round holes I envisioned. I put my hand on my scalp and found that the little dots didn't hurt when I touched them.

"How many more holes do they have to drill today?" I asked.

"There'll be no more. You've suffered the worst and it appears to be from what Dr. Simson has said, a total and complete success. Congratulations, Christopher."

I smiled but said nothing.

"Here," she said, "put these slippers on-you're going back up to your residence."

As I stood up to put the slippers on, I saw Dr. Karen Simson waiting beside a door. "I suppose you're here to take me back to my apartment."

"That's exactly what I'm here for," she answered.

When we entered my new quarters, I immediately walked into the dining room, sat down, and pushed the button that summons the chef. I was somewhat surprised to see that Karen was sitting on the other side of the table. I wasn't aware that she was still here-I thought that she had left.

"We have a few things to talk about," she said, and just as she said that, Henry the chef appeared through the double doors.

"How may I help you, sir?"

"Henry, I would like the best goddamn breakfast you can make," I appealed.

"And what might that be, sir?"

"Eggs, ham, bacon, sausages, steak, hash browns, orange juice, milk and whole-wheat toast; and don't forget the coffee," I exclaimed, and as Henry was excusing himself, I was already on my way to get a cigarette. When I returned, my coffee was waiting on the table. I took a big sip and a long drag of smoke. After another sip and another drag, I turned my attention to Karen.

"So, Dr. Simson, what other horrors do you have in store for me?"

"You seem to have a heavy appetite; and I mean that in a good way," she said, avoiding my question.

"Okay-to the point-what is it you want to say?" I asked, showing my impatience and arrogance.

"I'm trying to be cordial, Christopher, but if you'd like, I can be just as goddamn mean as you want me to be, pinhead!" she said while getting up.

"Okay . . . okay, Karen. I'm sorry, forgive me, but I'm ravenously hungry, bewildered, somewhat worried, and plum full of uncertainty. Is this really going to work?"

Karen sat back down in the chair and said, "Christopher, not only can I assure you that this transformation is going to work, but I can also assure you that what you will become, will astonish you beyond your wildest fantasies, and even your own perception of reality. It's my belief, at this time, that you just don't have the slightly insight of what you will become."

As Karen was ending her prophesying, Henry entered the room with my food. The smell was so exquisite that I wasted no time, and started digging-in while she continued to tell me what was to happen next. I ate like a starved nomadic, while she talked to the top of my red-spotted head.

"Eat up, Christopher, because this will be your last meal for about a week," and when she said that, I dropped my fork and knife on my plate, raised my red-spotted head and said, "Excuse me?"

"This will be your last meal for about a week," she repeated unemotionally.

"I don't understand; no food for a week? Can a person do that without starving to death?"

"Well, actually . . . yes. But I wouldn't recommend it. Please, Christopher, continue with your breakfast and I'll explain."

I picked up my fork and knife and did as she said. Before she started to speak, a thought came to me. This shit is like a series of poker hands. To my consternation, it seems to me that every time I get over a good deal, Karen plays a wild card, which changes the game. If it weren't for her assurances, I'd swear I was losing.

"Tomorrow we start the last phase. Although it's a long one, it'll be the last series of injections you'll need. We'll start again, at nine in the morning. You'll take another sanitizing shower, and then we'll proceed to the O.R. where you will be put into an induced coma for about seven days. We will feed you all the necessary nourishment intravenously throughout the week.

Everyday, I will be injecting newly mutated genes into all your bodily parts: your lungs, heart, kidneys, bones, liver, spleen, glands, muscles, skin, eyes, and everything else. Nothing will be missed, but it's a very time intensive endeavor, and so I decided to put you asleep. I will also promise you this; there won't be any pain. I can also assure you, Christopher that you'd swear it all happened in one day. You'll have no concept of time when you're asleep, and when you awake, it will have been about a week."

As I was listening to Karen I continued to eat, and after my stomach was stuffed, I called for Henry to tell him I was done. When the table was clean and Henry was gone, I poured another coffee and lit another cigarette. I now wanted to ask Karen some questions, which her comments hadn't addressed.

"Karen, when I'm asleep, will I have dreams? And if you say yes, will I remember them?"

"Indeed you will, Christopher, for the better and the worst. As far as remembering them, I honestly cannot say."

"This coma thing scares me-is it possible I could die in my sleep?"

"Absolutely not," she said, as I sensed a misleading remark.

"What about complications? What if something goes wrong? Will I become a vegetable . . . or will you terminate me?"

"Christopher, I promise you, nothing will go wrong," she insisted.

"That's what you say, but God only knows the truth."

"Christopher, the long and short of it is this: first, I'm telling you the truth. And second, we are at a pivotal point; if we do not finish the job, your brain will wreak havoc with your body which leads me to believe, that you will ultimately go insane, and I might also add, to your death. There's no stopping now, so we'll have to go on. Now, is there anything else you want to ask or say?"

"Just one."

"And what might that be?"

"Please, Karen . . . don't kill me," I said with embarrassing watery eyes.

And with that said, and to my surprise, Karen got up, walked around behind my chair and hugged me like a boyfriend. She kissed me on my cheek, then held my hand and said, "You're the bravest man I've ever met, and my feelings for you go deep. I promise you, Mr. Sandman, you're going to live and see your sweet Darlene again."

"Thank you, Karen," was all I could muster to say.

When Karen started to leave, she stopped for a moment and turned around. Her eyes were glassy and wide, and she looked at me as if it were the last time. "I'll see you tomorrow at nine." And with that, she turned back around and walked away.

The rest of the day was spent in thought-thoughts I couldn't ignore. I thought about Darlene, Billy and Scott, Bradly, Karen and Nancy and yes, even Dr. latex-Wayne Koloski. There were also the unknown masked faces and their mysterious eyes. I chained smoked and paced the floors. My emotions ran the gamut until I could barely take any more. All through the day, except another meal, I repeated this pattern until I happened to notice it was dark outside. I looked at the clock and realized it was time to sleep. I went into the bedroom and disrobed, laid down, took my pills and closed my eyes in hopes of some tranquil and pleasant dreams.

"Oh Sandman? Wake up sleepy head. It's the dawn of a new and wonderful day. Time to get up for your shower," I heard Bradly say.

"Where's Karen?" I asked.

"She's in the O.R.; preparing. Here . . . put these on," Bradly requested while tossing me a gown and slippers.

I followed the same routine as I did the other day with one exception. Before I took my shower, I was blessed with the gift of a super enema, and a good one at that. I shit like an elephant.

After I was all squeaky clean, Dr. Latex had me lie down on the table-bed. He inserted some IVs in me, and then began the journey to the elevator, which would eventually take me to the O.R. My vision started to become fuzzy and my hearing became strange with whispers, faint noises, bells and soft whistles-I even heard a merry-go-round. Dr. Latex must have given me some stuff to put me asleep. The last thing that I remembered was Darlene's pretty face.

"Christopher? Christopher . . . can you hear me?" a man's voice resonated hauntingly. Slowly, very slowly I opened my eyes. I shut them as quickly as I could. The light was too bright. "Open your eyes, Sandman, I know you're awake," the voice continued.

"I can't. It hurts too much," I complained.

"Try again," he said.

At first it was just a peek. Then I gradually opened them, little by little, until I could see without the light hurting me eyes. Scott stood beside my bed looking down at me. "How are you feeling?" he asked.

"I'm not sure. What's wrong with my eyes? Where am I?"

"You're in bed, and nothing is wrong with your eyes."

"What time is it?" I asked Scott.

"About seven."

"Well, Karen's supposed to come and get me at nine."

"It's done."

"What's done?"

"You're done. You came through it alive."

"Was there any doubt?"

"About fifty percent."

"Why, you rotten, lying, sons and daughters-of-bitches! You told me it was a sure thing!"

"And so it was-the injections, I mean. Bringing you out of that coma was the fifty percent chance."

"Okay. Now I'm pissed. Naked or not, I'm going to get out of this bed and kick your fucking . . . ouch! Ohooo, wow, sweet Jesus, oh damn that hurts!"

"You'd feel better if you'd quit moving around so much," Bradly said. I pulled the blanket down and looked at my chest. There goes those little red dots again, covering every square inch of my chest. And then I soon discovered, that my entire body was dotted in red.

"Are these from the injections that Karen was telling me about?"

"Yes," Scott said.

"How long before the pain goes away?"

"Right now," he said, holding a needle in his hand, "roll over, Sandman, this one's going in your ass. "There," he said, "give it about five or ten minutes and your pain will go away."

"Scott, why didn't you tell me that I only had a fifty percent chance-or for that matter, why didn't Karen? She lied to me!"

"We didn't want you to worry. Besides, it wouldn't have done you any good-we were going to do it anyway."

"Well, I'll be goddamned. Now how in the hell can I trust what you say?"

"Oh, quit your whining-ya big oaf. You're going to have to learn to trust me, I promise you no more holding back. Remember, Sandman, I'm still your friend.

All I did was withhold a little information so as not to cause you concern. I think you can get up now."

"Fifty percent," I said, murmuring under my breath.

"Why don't you get cleaned up, and when you're through, meet me on the ground floor," Scott said as he walked out of my bedroom.

When I went into the bathroom and glanced in the mirror, I became preoccupied with my reflection. My head had grown some hair, and as I looked closer, I also noticed that my eyebrows and eyelashes had started to grow back. My lips were dry and cracked, and my mouth tasted like foul paste, not to mention my putrid breath. My beard was as long as the hair on my head and it only seemed like yesterday, that I was as bald as a baby's butt. My chest was still hairless with the exception of some stubble. I hesitated for just a moment, and then let my eyeballs slowly fall down to peek at my genitalia. It was disconcerting to see little red dots on my dick and balls and everywhere else in between. I turned around quickly to see that there were more red dots on my ass. They were everywhere: My neck, face, nose and ears, arms and armpits and legs. Even some dots on my hands, between my fingers and also on my feet and toes. There must have been at least a thousand or more making me look like I had a fatal blow of acne.

After cleaning myself up, I got dressed and made my way to the elevator. I was not all that surprised that it took me to the ground floor where I met Scott talking to Sergeant Cooper.

"Well, here I am, your subservient tenant; what do we do now?" I asked Scott.

"We take a walk, Sandman, follow me."

"Where to?"

"Outside."

"Why?"

"To enjoy some sunshine and talk."

"Talk about what?" I asked as we walked through the doors.

"Your future."

"Damn! It's beautiful out here-I almost forgot how fresh the air smells. Tell me, Scott, how long has it been since I was last outside?"

"About three weeks."

"I didn't realize how warm it was. It seems like I've been . . ."

"Yeah, yeah, yeah, save it for yourself, Sandman, I have other more important things to say."

"So talk," I said as we aimlessly walked the grounds.

"We're going to keep you here for another couple of weeks. You'll be subjected to some tests and monitored by our staff to see how you improve."

"Improve? Oh, yes-to see how I change."

"Exactly."

"Then what . . . set me free back on the streets; back to my old apartment? Back to drinking and stealing? Back to my struggling days?"

"Not quite, Sandman, but in some way . . . yes. We'll turn you lose."

"Then what will I do?"

"You'll do what you told Darlene; you'll write."

"And what will I write about?"

"Anything you damn well please."

"Fiction?"

"If you want."

"Nonfiction?"

"Most likely."

"Like what?"

"Sandman, by the time you die, my guess is that not only will you have a huge collections of books, but you will also have written many sets of encyclopedias saved on computer chips."

"Get outta' here!"

"I'm very serious, Sandman. As a matter of fact, I'm prepared to give you an advance for the rights to your work."

"You're serious?"

"Does $125 billion sound serious enough?"

"Make it 150," I said jokingly.

"Okay. Oh, by the way . . . here are your new IDs I promised you."

Scott held out his hand and I took four cards: a drivers' license, social security, a Visa, and one to get me through the gate to this place. I took a few minutes to look at them.

"Scott, why do I need a drivers' license, you know I don't have a car?"

"You do now, here's your keypad."

"Three buttons?"

"One for your car, one for your home, and another for your front gate."

"Car? Home?"

"Sure, and they're all paid for-free and clear. Would you like to see your car?"

"I don't believe you."

"I'll show you-I'll even let you drive; can you handle a stick-shift?"

"I've driven everything from tricycles to tanks-you know where I grew up." We walked to a building that looked like a garage, and when we entered, I took a deep breath. It was a multiplex underground parking lot filled with fancy cars and trucks.

"Your car is over there," Scott said pointing.

"What is it?" I said as we walked closer.

"I chose this one just for you, Sandman. It's a vintage, modified bulletproof Jaguar XJR-9LM. Its top speed is 220 mph with a V12, and it only runs on octane 96. If you don't care for the color scheme, I'll have it changed."

"Good God, no! I love those colors of black and gold."

"It's all yours," Scott said as he opened the door to the drivers' seat. He held out his arm and hand like a game show host.

We got in and I fired her up. She had thunder under her hood, and when I stepped on the clutch and gas, she roared like a wild beast. Scott pushed a button on a remote and a door opened to let us out.

"Put your seat belt on, Sandman. I want you to drive to the city and please, keep it under 150 mph on the desolate roads."

As we made our way through the main gate, I punched the gas and made two donuts by mistake.

"Wow! Jesus Christ, Sandman, get a hold of this thing!"

"Goddamn! This bitch is great!"

All the way down through the mountainous roads, I kept it under 100 and I never did make it to the fifth gear. Finally, I slowed down to less than 60 mph when I got into some light traffic and saw the city get bigger. Scott pointed me to the left and the right, and other various directions until we ended on the west side of town.

"Okay, Sandman, turn down this road, and go slow," he said as we passed through a well-kept neighborhood with big, beautiful and fancy homes. The farther we went, the bigger and better they got.

"What's the name of this street?" I asked.

"Salisbury Lane. Now slow down and turn into this drive."

"All I see are trees."

"Keep going until you come to the gate."

I found myself driving slowly up a small road, covered by trees, bushes, hundreds of flowers and clean-shaven lawns until I finally stopped at a black, heavy stronghold gate. Scott showed me which button to push on the remote and the gates opened like prison doors. I drove through at a snail's pace until we reached a huge, three-story, cobblestone bastion.

"Is this my house?" I asked, now feeling like I had to shit.

"Well, that's part of it. That's the entrance; pull over there."

When we got out of my car, I found myself loping beside a gigantic, cobblestone structure. "Sandman! Come back here," I heard Scott call, "you've got the pad to get in and I don't."

The house was mammoth and beyond my belief, and as we entered the foyer, I heard my voice echo: "Holy shit!"

"So, what do you think?" Scott asked.

"It's terrific! But what is that home behind mine by those pine trees? Are they my neighbors?"

"No, they're not your neighbors. It's called a guesthouse for your friends and other people to live or stay temporarily."

"Wow! Wait until Darlene sees this," I response exultantly, while Scott shook his head approvingly.

"Let me show you around," he said.

Every room was unbelievable. The furniture looked antique and unpleasant to sit in, but was instead, extremely comfortable, and very valuable. And under all this self-important shit lay hidden secrets of uncommonly high technology.

The first floor had nay a window, but the upper two floors did. A self-contained, electrical generated computer concealed by a wall in the basement computerized everything that led to an outlet in the wall, and even some things that didn't. I could even talk to the house as if it were a person: "Turn the lights down-dimmer; that's good," or, "Close the curtain. Shut the door. Turn the TV on; change it to channel 12. Run me a bath-110 degrees Fahrenheit. Flush the toilet. Turn the sink on . . . warmer, warmer please, okay, that's fine," and so on. To prompt my home for a command, I had to address it by name, which was simple: "house." I also had the ability to turn the voice actuating system off and operate my home manually.

"One last thing before we leave, I want to show you your wall-safe," Scott said proudly.

It was down in the basement, which was divided into rooms and finished in luxury like the rest of my home.

"Follow me, Sandman, and I'll show you your office."

The room-my office-was simply amazing. It was lined with disk shelves containing a multitude of computer disks, and under the center of the three large windows, sat a computer. Adjacent to this wall was another wall embedded with TV screens and monitors for security. But something seemed strange about the room.

I felt like I was standing in a well-lit study-like the one that I saw on the second floor. I'm sure it was the windows that first struck me as odd.

"Scott, how can there be windows on *this* wall where the computer is, when I know for a fact that there's another room behind it?" I asked.

"These are not real windows, and neither is the sunlight that's coming through. It's an illusion. They're really ultrahigh-definition, backlit-plasma screens. But make no mistake; the images you're seeing are real. You're looking at three beautiful views of your surrounding landscape; but you're welcome to change them if you wish. Do you feel like you're standing in a basement?" Scott asked with his hands in the air.

"No, I don't."

"Then the job was well done. Your office was designed not only for comfort, but also for your safety-you're standing in a shelter. It even processes its own supply of air."

"You think I need all of this shit, Scott?"

"Just a precaution, Sandman. Now, back to your wall-safe. Do you think you can find it?" Scott asked.

After spending a few minutes fumbling around the only wall that had no backlit, plasma-screens or monitors and displays, I told Scott, "I give up."

"Speak, instead of looking, Sandman."

After Scott said that, it took me less then a second to figure it out. "House: open the safe," I said out loud. And with that said, a section of the wall slid sideways while all the time I was standing right in front of it.

"Now push the red button on the front of the steel doors," Scott instructed. The thick steel doors parted for Scott to gently push me in, while he followed. The lights automatically came on.

"Here is the deed to your home and the title to your car. Here are your banking documents, including your checkbook. Over there, on that shelf, is a quarter million in cash. And over there, for keepsake, is your old .38. Over here is a pantry with food, water and other supplies. There's more here for you to look at, but I'll let you do that on your own. Any questions?"

"Could I accidentally get . . . ?"

"Locked in?" Scott said, finishing my question, "Absolutely not! I'll let you demonstrate. Tell the doors to close," he said.

"House: close the doors," I commanded, and indeed they did.

"Now look over here in this monitor and tell me what you see."

"I see my office and the wall-safe closed. Just as it appeared when we first walked in. But obviously it looks empty because we're in the safe."

"Very well. Now if you don't mind, it's time to go, Sandman. I have an appointment. Open the doors."

"No. You," I said.

"I can't, Sandman, only you can."

"Say the command!" I insisted.

"It won't work for me."

"Say it!"

"Okay, goddamn it. House: open the doors," he commanded and nothing happened. "You see?" he said convincing me.

I said the same command and the doors opened. When we left the confines of the safe, Scott started to yell at me. "Don't you ever, ever try another stupid stunt like that, and never question my authority!"

"Sorry, Scott. You're a very intelligent and wise man. I'm sorry. I've made you angry and I sincerely apologize.

Please, just let me say, I will never, ever, turn on your words again. You've endowed me with life and even more, and from this day forth, I promise you, I'll be at your everlasting command-will you forgive me, Scott?"

"We'll forgive each other on this day, never to be thought of, or said again. Now, let's get back to the Society."

As the days went on at the Society, I found myself in the library reading books and writing on the computer. I first read fiction, then fact. I was also drawn to the chemistry and physics labs they have. I even redid some famous experiments, and a few of my own. But within this time, I was also being tested on everything my body was doing.

The red dots finally went away. My hair grew back; my skin became thicker and yet, strangely more sensitive. My I.Q. went up sizably; my eyes could see sharper and farther; my height grew a couple of inches and my strength grew ten fold-I could even run faster and jump higher than ever before. I was also immune to simple and deadly diseases among many other attributes. The staff at the Society called me a genetic neophyte-and now they were setting me free.

CHAPTER THIRTEEN

"Sandman. Before you go, I want to give you a little protection."

"What caliber is this?"

"Slightly greater than a .457. But it doesn't use bullets, nor does it need a silencer. It works on the principles of coherent radiation and atomic batteries. Do you know what a *maser* is?"

"It's a device that works like a laser-but instead of using visible light; it uses transparent microwaves. The same wave-lengths used in microwave ovens."

"You must be reading, Sandman-you're becoming much more adept-I like that new quality in you. Maser is an acronym for microwave amplification by stimulated emission of radiation.

Here's the shoulder holster. Now don't let this little toy fool you-it has the power to punch a hole through your head from the distance of a football field. Or you can set the power on low; which is non-kill. And as I just said, there's the added advantage of no shells, and no sound."

"A gun that shoots deadly, silent invisible bullets. That's pretty neat, but I thought that's already been done?"

"Of course it has, but never in a device which you now hold in your hand. I demand that you take some time and practice with it, and always wear it wherever you go."

"Scott, why are you giving me this?"

"I've already told you; for protection. I hope you only use it in self-defense, but essentially, and regrettably, I'm giving you the license to kill. I have to. You're a very valuable asset not only to yourself and us, but also to the future of the generations to come. And that, my friend, lies distantly in front of you.

Also, and never forget, we'll be keeping in touch. If you ever need anything, here's a number to call-memorize it and tell no one. Not the number, this place, or who you are. From this moment on, you're sworn to secrecy. I don't what to get into the ramifications if you should break this rule-you know how I feel about you. So now, Mr. Christopher Sandman Lance, you're free to go."

After my farewell, I got in my car and sped away, driving like a lunatic to see Darlene. I pulled in her driveway next to a strange car and rang the doorbell. She opened the door just a crack, but wide enough for me to see, by her mishap, a strange man trying to stand away from this noteworthy scene.

"Hello, Darlene. I'm back. I came over to see you, but I can see you're busy seeing somebody else. Have a nice life," I said while turning around and walking away.

"Christopher? Christopher . . . please don't go away. Come back and let me explain."

I stopped, turned around and saw her pleading green eyes, while she held the door wide for me to come in. Her expression convinced me to walk back, and through the door I stepped. "Christopher, this is my boss, Mr. Sam . . ."

"I don't give a fuck who he is-why is he here with you in a negligee. It's three o'clock in the afternoon for God's sake!

Explain that to me?" I angrily told her while staring like fire into the man's eyes, and seeing his knee's starting to shake.

"Christopher, I've been sick for the last few days, and Sam just stopped by to drop off some manuscripts for me to read. That's all there is to it," she begged while swearing the truth to me.

I touched her shoulder and moved her around, and then told her to look Sam in the eyes. I gently held her hand, put my chest against her back, and with my other arm, I wrapped it around her chest-just below her breasts. I then bent my ear close to her mouth. I was preparing a test. With my new acute senses, I would be able to tell if she were lying, or telling the truth. I could feel the moisture on her hand-her heart beating and her breathing. Her voice and inflections were very acute to me. I could also see her eyes in a mirror. I was testing my skills as a human lie detector.

"What are you doing, Christopher?"

"Just stand the way you are, and answer some questions for me."

"Okay, Christopher, but this seems very strange," she told me while looking Sam in the eyes.

"What is your name?"

"Darlene. But you know that, why are you asking me my name?"

"Just answer my questions and nothing more. Now . . . let's start over. What is your name?"

"Darlene."

"Did you eat earlier today?"

"Yes."

"Do you work at Martin & Kelly Publishing?"

"Yes."

"Are you a virgin?"

"No! Christopher, why are you doing this to me?"

"Please, Darlene, just answer my questions."

Although she wasn't aware that I was calibrating my senses to hers, I now felt confident enough to see if she would tell me a lie. And all the while, Sam stood looking at us not quite believing what he was seeing.

"Is the man standing in front of you named, Sam?"

"Yes," she said, telling the truth.

"Is he your boss?"

"Yes," again I sensed a truthful answer.

"Did Sam come over to drop off some manuscripts?"

"Yes," was another truthful answer.

"Is he your boyfriend?"

"No," another truthful answer.

"Is he your lover?"

"Hell no! Now let go of me. I've heard just about enough of this shit!" was another truthful but angry answer. I let go of Darlene and stepped back away. I felt it. Her breathing, her voice, her heart rate and perspiration remained the same. I now knew it-she was indeed, telling me the truth.

"May I ask just one last question?" I said politely.

"Go ahead," she said, looking scornfully at me.

"Why are you wearing a negligee in front of your boss?"

"He just arrived, and I forgot to put my fucking robe on! Okay?"

"Oh my. I feel so embarrassed. Please except my most sincere apologies to you, sir. And to you, Darlene, will you ever forgive me? Please? I've been away and have missed you deeply. I'm sorry I let my emotions get away.

I've behaved in the manner of a total fool. Please continue with your work, and I'll just be on my way."

I stepped out the door feeling kind of foolish, but along with this feeling was the over powering knowledge, that I discovered how to tell if someone were telling the truth, or lying. I was halfway to my car when I heard Darlene's voice say, as she leaned out her front door: "Christopher, please call me tomorrow?"

"What time?" I asked.

"The sooner the better," she said with a questioning smile.

I smiled back in a simple way, then got in my car and drove away.

It wasn't the first time, but I found it exciting to see people point at me and my car.

I knew they were impressed, and maybe even a little envious; but I really didn't give a shit. As far as I was concerned, I served my time: in the destitute ghettos, the raunchy apartments, the vomit-stinking hallways, not to forget the bums and derelicts including the gangs and the thugs and the unlit back streets. Fuck them all and there damned lives-I'm on a new pathway.

As I was driving home, Billy came into my thoughts, and so I decided to turn around to go see him. When I pulled in front of his store, I pushed a button to electrify my car. Scott had shown me this little trick. If someone should try and touch my car, they would receive about 10,000 volts of electricity. Of course, not to kill or cripple-the wattage is too low for that; but to simply shock them away; like a farmer's electrified fence. However, the computer in my car is capable of distinguishing between a child's finger and that of an adult. The children are safe, but the adults are not.

When I walked in the store, I was surprised to see Billy sitting in a wheel chair, looking haggard and older then he should. The store's inventory was low and his shelves were dusty with old items and expired expiration dates.

"Hi, Billy, it's me, your friend Sandman," I smiled. But there wasn't the slightest hint of joy or happiness at seeing me. I suddenly felt a very strong sense of sadness and concern.

"What has happened to you in the last two months while I was away?"

"Please leave, Mr. Sandman, my business is failing and so is my health, my wife, Betty is dead-there was no one around who would help me."

"Oh . . . I'm sorry, I didn't know, but please don't ask me to go. Billy, I'm your friend. Tell me what happened; maybe I can help. Please, Billy, don't shut me out, please . . . tell me what happened?"

"I was robbed again. Betty tried to shield me as the bullet when through her heart and into my spine. They got away with five dollars. Now, the only money I make is from selling my liquor, beer and wine, including the cigarettes. I've been drinking up most of my profits and eating the can food that's left on the shelves. I can no longer pay the rent, and the health inspectors are forcing me to close my fucking store.

Thank God, I still have my gun, which I'll use on myself when the authorities come to pad-lock my doors and throw me out on the streets."

"I'm very sorry for what's happened to you and your wife, Billy. But I'm in a position to help you now."

"How? By paying me your liquor debts?"

"More than that, Billy. How would you like a new store, with managers, stockers, cash registers, checkers and other store clerks? We can set it up on the west end of town where people have money."

"Please, Sandman, don't fuck with me, I'm in no mood for levity."

"How long before they close this dump?"

"About a week; maybe a little more."

"Do you have *any* money?"

"Well . . . the thieves stole my cash register after we were shot. But I have about sixty dollars and some change in this shoebox."

Billy, I've got to get you out of here and some place that's new. Your life has been hard and it's time to move."

"Sandman, I really do appreciate your thoughts, but as you can see for yourself, I'm at the end of my rope. There's nothing more I can do but simply let go."

"Well, my fine friend, I'm throwing you a new life-line and you'd better take hold. I'll find you a new store, but I'll need a few days-can you wait that long?"

"Sandman; quit talking out your ass, and please, just go and leave me alone."

I reached for my wallet, which Scott had given to me as another going-away present but I had never opened. I was hoping to find a little cash in it and was flabbergasted to see and count, five thousand dollars including a note: "*Just a little chump-change, my son. Be safe-your father, Scott.*"

"Sixty dollars is all you have?" I asked Billy as he rubbed his swollen red eyes.

"I have some good scotch, if you can afford to buy?" he asked me sheepishly.

"Fuck the scotch; here's two grand. Will that get you by for a few days?"

"Oh my God, Mother of our Lord Jesus, you've crossed the line. I've prayed for you that it wouldn't be this way. And how many people have you killed? Forget it-I don't want to know. Take your dirty fuckin' money away, because I don't want it. Are you that stupid and blind! Look at me, Sandman; look at what bad people-and now you-have done to my wife and me! Take your goddamn money and go, I will be deceived no more!" Billy scolded me, crying like a baby and forcing his words.

"But, Billy? I'm telling you the truth. I'll buy you a new store, and the money I have, came honestly. I've made it through a contract for my writings. Please, Billy, just give me a chance, and someday soon, I'll show you proof."

"Just go, Sandman, and take that damned money."

Sadly and slowly I walked out of the store, but left the money on the counter beside his shoebox. I pushed the keypad to unlock the car door, which also disengages the electricity. All the way home, I couldn't help but to think about poor Billy. Life really dealt him a shitty hand and I couldn't get him out of my mind. I felt sorry for him, and who wouldn't? I was determined to pull him out of that God-forsaken situation. Even in my earlier times of desperate needs, he was always there to help me.

The very first thing I did when I arrived home was to call that special number that Scott had told me to memorize. It rang one time before I heard Sergeant Cooper's voice on the other end.

"How may I help you Mr. Sandman?"

"I need to talk to Scott."

"I'm sorry, Sandman, but Scott is not available at this time. Let me transfer your call to Bradly."

"Hello, Sandman, what can I do for you?" Bradly said in a subservient voice.

"Bradly, I need a lawyer and a commercial real estate agent."

"Is there a problem?" he asked sincerely.

"Well . . . yes and no."

"Could you be more specific?"

"I want to help an old friend, Bradly, let me explain," and I proceeded to tell him the story about Billy. And I must admit, I knew that from now on I'd treat Bradly with much more respect. He was completely sympathetic with all that I said and before he hung up, he asked me to wait by the phone. Within ten minutes, my phone rang with Bradly on the other end again.

"Okay, Christopher, I have an appointment set up for you at the *Emerald Hotel* at ten in the morning tomorrow; conference room 503. The lawyer's name is *Dan Handson* and the real estate agent's name is *David Frances*. They will help you solve your friend's problem. These guys are top notch, so listen to them carefully. If you need anything else, never hesitate to call me. Are the arrangements satisfactory?"

"Perfectly. Thank you very much, Bradly."

"You bet," Bradly said before he hung up.

I felt relieved and less distressed now that I had the wheels in motion. Billy would get a new store, bigger and better than the one that was going into foreclosure. Although he didn't believe me, I knew his life was going to get better. I also wanted to help with his medical condition. That old wheel chair that he was sitting in looked like it was about ready to fall apart.

As I sat thinking about Billy, I decided to add one more final goal: to see if there might be an operation for him so he would be able to walk again, but that would have to come later.

Looking at my new *Rolex,* I saw that it was going on seven. This was the first night in my new home and I felt like looking around at things just a little bit closer. It occurred to me as I walked from room to room, that there were no pictures.

I also noticed that there were no knickknacks or other displays-nothing of interest that was personally mine. But that was to be expected-I didn't have any. I decided that this would change in the near future. I needed to personalize my home with things that pleased me.

Eventually, I found myself in the kitchen and discovered that I was hungry. As I spied around the cupboards, drawers, pantry and refrigerator, I found that there was food galore. Frozen food, canned food, fresh vegetables, assorted sandwich meats and other delightful treats. For the first time in a long time, I couldn't decide what to eat. Feeling a bit lazy, I pulled a frozen meal from the freezer and sat it on the island counter. "Home: preheat the oven to 350 degrees." It really didn't surprise me to see the oven light come on while the heating coils started to glow red. "Preheat complete," the oven's female voice spoke back. I placed my dinner in the oven and then said, "Home: bake my dinner for 45 minutes, then set the temperature on warm."

As my dinner was baking, I went to a room that I skipped over to explore. It was directly behind my garage. When I opened the double French doors, I was dumbfounded to see an Olympic size pool with a high dive and low dive springboards. The shallow end was three feet while the diving board end was twenty-five feet deep.

After eating, I decided to take a swim and found myself spending hours on end, swimming and diving off the high dive. Of course, I wore no swimming shorts because I didn't need them. After the hours of play, I got out of the pool for the night. I put on a robe and heading for my living room to relax and have a drink. I kicked-back in a comfortable chair and began watching the television.

After about six drinks and half a pack of cigarettes, I called it a night. The master bedroom suite was on the third floor and I told the house to wake me by seven. I laid in bed and read a book until my eyes became droopy. I put the book down on my nightstand and told the house to turn off my room lights and with a little passage of time, I was dreaming.

I heard a voice-that female voice telling me to wake up. "Time to get up, Christopher, it's seven in the morning," it spoke gently but loud enough to stir me awake. Remembering I had things to do, I followed the voice's command. I got out of bed and headed for my bathroom that was fashioned much like that at the fourth floor apartment at the Society. After cleaning up, I went down to the kitchen to make myself some scramble eggs and toast.

By nine-thirty, I was in my car and headed for the *Emerald Hotel* to meet with the attorney and real estate agent. At ten to ten, I was by myself, sitting and waiting in conference room 503 at a large and fancy table. Just about a minute or two before ten o'clock, two men walked in and introduced themselves as we shook hands.

"Good-morning, Mr. Lance, my name is Dan Handson, your attorney from *Segal & Zales*. It's nice to meet you," he said while shaking my hand.

"And I'm agent David Frances, from *Frances Commercial Sales*. I understand you're looking for some property," he said with a professional, but pleasant southern dialect.

"Please, have a seat gentlemen, and I'll explain why we're here," I said with a sophisticated attitude and voice. I sat at the table with nothing, but they on the other hand, place their attaché cases on the table and pulled out some papers, note pads and other assorted items. Each one sat on opposite sides to my left and right.

"I'll start first," David said in that southern accent, "I understand you're looking for some property that would serve as a super market food store on the west side of town."

"That's right," I said, "and I'm looking for a high probability of traffic exposure to help ensure a successful start. Preferably, places on the corner of two busy main roads with easy ingress and egress from both sides, and plenty of parking space. Do you know of any such buildings on the market?"

"Actually, I do," David said as he handed me a set of pictures from different locations, "Here are a few prime localities that I downloaded from the Multiple Listing Service yesterday after talking to Mr. Bradly Hues."

"Hum . . . I like this one the best; what do you think?" I asked.

"Very good choice, Mr. Lance, that would be the one I thought you'd choose; would you like to submit your bid today?"

"How much?"

"The property was appraised at 1.2 million. The latest submitted bid stands at 1.25 mil, and the closing date is tomorrow at noon," David replied.

"Is there enough square footage?"

"Plenty, and then some. It also meets all the requirements and zoning laws established by the state and federal government. It's a perfect piece of real estate and most importantly, it's not encroached by any other food stores; it's a very good location," David said, invoking my excitement.

"Are there any drawings?" I asked.

"Yes, but they're in my office. I have the architectural and civil engineering plans from when the building was established, and nothing's been changed," he added.

"Double the bid and submit it to your company; I'll want to start work on it as soon as possibly."

"But, and with all due respects, sir, I think you'd be paying too much."

"Are you telling me I'd be paying too much; what's your stake in this?" I asked.

"My commission is two percent, but I'm an honest man and my job also includes your interest and success; you needn't pay over the fair market value, sir. I am a very well respected professional in my field, and I don't over submit my client's monies. I can guarantee you that 2 mil will buy you that store; lock, stock and barrel."

"A true professional. How nice. I enjoy working with people like you; a man with integrity *and* money; how very rare," I told David, as if this was something I do everyday.

"Very good, Mr. Lance, now if you'll just sign these . . ."

"Give the documents to Mr. Handson for his scrutiny. And while we're at it, Mr. Handson, I'll want to transfer the deed to a friend of mine with the inclusion of 10 percent of the net profits going to me," I said.

"Very well, sir, but I'll need some information . . ."

"You'll get all the information from my friend today, and I'll also need your expertise to help extradite my friend, Mr. William (Billy) Richers, from a legal matter with his existing property and financial hardships," I interjected.

"When and where do we meet this gentleman today, Mr. Lance?" Dan asked.

"Let's set the time at three o'clock in this room; would that be a problem with either of you gentlemen?" I asked.

"No problem," David answered.

"I'll be here," Dan followed David's answer.

"Great. Three o'clock it'll be. And by the way, gentlemen, let's work on a first name basis. Okay?" I asked as they both agreed. "Very well, now if you'll excuse me, I have to retrieve my friend, Billy, to meet you two at three."

On my way to Billy's, it suddenly occurred to me that I had completely forgotten to call Darlene. I figured I'd call her from Billy's store, hoping she'd be home.

When I arrived at Billy's about one-thirty and walked into his store, I saw him sitting in his dilapidated wheel chair, next to the counter with his money shoebox, and a blanket covering his lap. He watched me enter.

"Now what the hell are you doing here again? Did you come back to get your money; well, it's too late, I've already spent it," he scolded me.

As I looked around, I saw that nothing had changed. The shelves were a little emptier, but still covered with dust. There was mud and grime on the old, unpolished wooden floors, and nasty dribbles and smudges on all the glass. To sum it all up, I'd never realized just how depressing and dirty this place really was. I couldn't believe that the health authorities hadn't chain-locked his doors well before this day.

"Billy, I want you to come with me. I have a surprise for you-it'll change your whole life for the good. You deserve much better than this place. Please, just give me a chance-trust me, Billy, I'd never do anything to hurt you," I pleaded and begged.

Billy looked up and surveyed my eyes; "Fuck you!"

"Why are you so angry at me? I've never done anything bad to you," I replied in a stern tone.

"It's people like you who contrived my demise! You are a murderer! A gun toting thief; a lair and bum! I know you well, Sandman, you've got something up your sleeve.

You're trying to connive me, or deceive or tick me into one of your schemes-and I'll have no part of it! Go now, Sandman, just go the hell away," Billy cried sorrowfully.

"I will not leave you here in this shit hole! You're coming with me even if I have to carry you myself!" I demanded.

"You'll do no such thing!" Billy screamed.

"The hell I won't; just watch me," I said with determination in my voice as I snatched him out of his wheel chair.

"Put me down you bastard or I'll call the police!" he yelled as he squirmed in my arms. I suddenly realized just how strong I was, because Billy felt as light as a feather. I placed him in my passenger seat and closed the door quickly. As I was walking around the car, I could see, but not hear him, screaming and pounding his fists. When I opened my door, I was basted by his noise-his screaming and thrashing.

"Shut up!" I yelled and started the car, "just shut the hell up!" I fastened my seat belt and shifted into first gear, laid on the gas pedal doing donuts in the middle of the street. I had to laugh because of the smoke and bouncing tires scared the shit out of him as he grabbed onto the seat. He looked at me wide-eyed with his mouth making an O and from the expression shown on his face, I'm sure he thought I was quite insane.

I straighten the car and made a beeline with the engine roaring and the tires screeching as I shifted through the gears. I stopped accelerating when I reach just a little over a 120 miles per hour, running a red light and risking it all.

"Slow down! Slow down!" he shrieked.

"Only if you'll shut up and fasten your seat belt," I yelped.

"Deal! Deal, you've gotta deal! I'll shut up! Just slow this damn car down!"

I slowed the car to 45 mph and said, "Will you now, please, just hear me out?"

"I'm listening, I'm listening, Sandman, you've frightened me; where are you taking me and why?"

Pushing an un-pushed button I've never pushed before, a holographic GPS screen displayed itself.

"Quiet for a minute, Billy," I said as it suddenly hit me; "I had a phone in my car." I started talking to the phone speaker and telling it to dial Darlene's number while I recited the digits, and then hearing her phone ring.

"Hello?" she answered.

"I'm sorry, Darlene, I forgot to call you earlier. Something came up. An important business deal that I had to take care of-will you forgive me again?"

"Christopher! You called me?"

"You thought I wouldn't?"

"I wasn't sure."

"Can I see you later?" I asked.

"Of course, I'd like that; I've missed you, Christopher."

"And I miss you. Is seven or eight okay?"

"Perfect. I'll make us dinner, would that be all right?"

"Sounds nice, I'll see you later."

I looked over at Billy sheepishly, "You remember Darlene; don't you?"

"I remember her name, but I don't know her," he said after exchanging glances with the phone and me and sporting a look of wonderment. "What year is this? What are you doing with this incredible car? I must be in a dream; some kind of fuckin' la-la land?"

"You're not dreaming, Billy, I'm rich-I'm a multibillionaire-can you believe it!"

"Is this for real; are you for real?" he said, still not sure of his surroundings.

"You'd better believe it. I was trying to tell you in the store, I have a surprise for you."

"What surprise?" he asked unsure.

"Your new store!" I bellowed and laughed.

"You weren't just joking?"

"If I were joking, then where did this car come from, and the two thousand I gave you?"

"I don't know, maybe you robbed and killed someone," he said quite honestly.

For some unknown reason, I lost control of my emotions and went temporarily berserk. I slammed on the brakes and the car threw us forward. The sudden deceleration pinned us to the seat belts while the tires smoked and screeched to a stop.

"Now listen to me, and hear me loud, I didn't steal any goddamn money; I earned it! And if you only knew what I went through, you still wouldn't even believe me! I told you earlier, I'd prove it to you later. But, if you continue to doubt my word, I swear to God, I'll turn this thing around and take you back to that rotten hole of a store you call home. I'm trying to help you Billy, quit doubting me. Now; I'll let you decide: do we go back or forward?" I yelled angrily.

"Okay, Sandman, I see you're serious, so let's go forward," he said apologetically.

I shifted back into first gear, and accelerated slowing while obeying the speed limits and stoplights. In a calmer voice I said, "We have a three o'clock appointment at the Emerald Hotel with my attorney and real estate agent. I'd like you to look at some pictures and talk with my attorney; are you with me?"

"Yes, I'm with you, Sandman," he surrendered.

"Good. By the way, Billy, you stink. When was the last time you took a shower?"

"I don't know, Sandman, maybe a week ago," he said shamefully and sadly.

"Well, that'll change when we get to the hotel; it's just around the corner." I avoided the valet and found a spot to park near the entrance. We had a little discussion.

"I don't have my wheel chair, Sandman, how am I suppose to get around?"

"I'll carry you."

"You've gotta' be kidding. You know how stupid that'll look? I refuse to have you carry me in your arms like I'm some sort of a little child! I'd rather crawl."

"Now, Billy, don't fight me on this; we'll get a wheel chair when we get into the lobby."

"All right, okay. Geeezzzz! This is going to be so embarrassing," Billy complained.

After I picked him out of my car, I started for the lobby, and all the while, Billy hid his face in my shoulder. I had to smile a big grin because it did indeed look ridiculous. We looked like a couple of fagot lovers, who were going to check in so we could spend our honeymoon taking turns fucking each other in the ass. There were a few people who stopped and watched and some of them even laughed.

"Nobody's even looking at us," I told Billy as a little girl pulled on her father's sleeve and pointed.

"Don't tell me nobody's looking, I can hear them laughing," Billy whined.

When we were in the lobby, I immediately went to the desk, then sat Billy on the floor for show and yelled, "Can we get a damn wheel chair!"

The desk clerk looked at us strangely. I reached in my wallet and pulled out one thousand and placed it on the corner.

"Quit looking at him-look at me, I need a room and a wheel chair," I demanded.

"Yes, sir, right away, sir, the wheel chair will be here in a moment," he said snapping his fingers.

"What are your rates," I asked the clerk.

"For the room?" he asked.

"No, for your whores in the lobby! Of course for the room; what are you; stupid! I said loudly.

The clerk stiffened up and cleared his throat and said, "That all depends, what type of room are you looking for?"

"Well, for starters, one that you can sleep and shower in, and also accommodations for the handicap," I said sarcastically.

"Sir, you needn't be so abrupt, all of are rooms are designed for the handicap," the clerk said disquietly.

"I apologize, my good man, please forgive me, but our time is running late. I need a room with a single bed and an alcohol cabinet; I trust you do have such a room?"

Most certainly, sir; the room cost $350 a night," he said politely.

I dug out another $1,450 more from my wallet and added it to the one thousand setting on the counter, "Here's $2,450; can I have the room for a week?"

"You most certainly may. Oh, and here comes your wheel chair. If you need some . . ."

"I'll take care of my friend while you prepare my initial receipt thank you very much."

Billy's room was on the fifty-third floor, and a very nice room it was. I helped Billy take off his clothes so he could take a bath-it was almost one o'clock.

I hollered threw the door, "Tell me when you're through and I'll turn the shower on so you can rinse, shave and brush your teeth. Oh, and another thing, what size is your waist and feet?"

"Forty-three and ten D," he yelled back at me, "and I'm almost through with my bath."

I handed him a shaver and a toothbrush with a tube of paste, which I gathered from the collection of amenities displayed on a bathroom shelf. As he was shaving and brushing, I called the front desk to have them bring up a suit and a pair of shoes; included socks and underwear. Then I dialed for room service.

"Room service," a voice answered.

"Could you hold on for just a second please," I asked, and then yelled at Billy, "Hey, Billy, you like lobster?"

"I love it! But it's been so . . ."

"Could you have two lobster plates brought up to room 5342?"

After I hung up the phone, Billy suddenly called out, "I'm done, Sandman, could you help me get out of this fucking tub?"

Forty minutes later we were sitting by the panoramic view of the city, eating lobster and dressed like businessmen. The food was great!

We had about twenty minutes before the meeting. While we were eating, I explained to Billy that he would be staying here for a week. Of course, it was okay with him. I also explained the purpose of our meeting and told him the names of the two men we were going to meet. Intermixed with our conversations, I also included some details of the case. I was pleased that he took me seriously and also listened carefully.

After eating, we took turns inspecting ourselves to make sure there was no mess on our shirts and ties. It wasn't long after that when we found ourselves sitting in conference room 503 and waiting. I gave Billy a little vote of confidence with a wink when the two men walked in. I stood up to exchange handshakes again, while Billy sat and looked.

CHAPTER FOURTEEN

"Gentlemen," I said, "I'd like to introduce you to my very special friend, Mr. William (Billy) Richers."

"Very nice to meet you, Mr. Richers. My name is David Frances; your real estate agent," he said in that very pleasant southern dialect.

"And I'm attorney, Dan Handson, Mr. Richers; it's a pleasure to meet you."

After all the handshakes, we sat down to do business.

"Okay. As I asked you two gentlemen before, and for time sake, let's keep this on a first name basis. Is that okay with you, Billy?" I asked.

"Fine with me, Sandman," Billy said. And when Billy said, *Sandman*, the other two looked at each other in a bewildering and questioning glance.

"Ah . . . that's my nickname, gentlemen. Billy always calls me that," I responded straight-faced.

"Let's start with you, David, I'd like you to show Billy the pictures, and I trust, some drawings of the building I've chosen."

"No problem, Christopher," he said as he pulled the pictures and drawings out of his briefcase. "As you can see, Billy, the locate . . ."

"Mr. Sandman?" Billy interrupted as he looked at me.

"Ah . . . yeah, Billy, but please don't use my nickname," I told him while giving him the *evil-look*, making sure the other two didn't notice.

"But you never told me your real name," Billy said with a questioning expression that I almost took for anger.

"Yes, I know, Billy, but you've always called me by my nick . . ."

"So, Christopher, what is your *last* name?" Billy requested.

"Ah . . . listen, gentlemen," the attorney said, "I sense a bit of confusion between you two. This may cause some problems; legal problems if . . ."

I didn't give the attorney time to continue, but simply spoke instead. "Gentlemen, I assure you, there are no problems here. Billy has always known me as Sandman; I've never told him my last name. All my friends call me Sandman; it's been that way from childhood. It's nothing to be concerned with; it's just a silly nickname my parents gave me. When I first met Billy that was the name I used; and it's never changed. Now, before we go off on any more tangents, let's say we get down to business. Please continue, David."

David started to gather his pictures and drawing, while Dan was focusing his attention on the legal papers and putting them in order. Both their eyes were on the table, while Billy's eyes were on mine. I took a momentary second and pressed my finger against my puckered lips, hoping that Billy would get the hint. Hoping he would understand my signal, which I saw he did. Then quickly looking back at the other two, I knew they didn't see the clue-they were still busy with their pictures and papers.

"Okay, Billy," David continued, "Here are the pictures and some architectural floor plans. As you can see by the dimensions, there's plenty of square footage."

I could see Billy's eyes light up when he finally understood the seriousness of the two men talking.

He would glance up at me from time to time with a big smile on his mug, which in turn, put a smile on mine. After the real estate agent was through talking, the attorney took his turn. When all was said and done, we had a business deal. Only three hours went by before we were all standing up, except Billy, and shaking hands. It was the old cliché: a done deal.

When we got back to the hotel room, the criticism started. I knew that Billy was full on confusion; his anger was mixed with joy, but his anger came first. And I could also tell by his words and expressions that he was on the edge of an emotional cliff and I didn't want him to let go, but yet he did.

"Sandman-goddamn it all, tell me what the hell's going on! A month and a half ago you were roaming the streets for money and now you're buying a store. I demand to know where you got all this fuckin' money! You're supposed to be my friend, but friends don't lie and hide secrets-or do they? All these years I've been calling you Sandman, and you told me you wanted it that way. But now I don't know what to believe-hell, I don't even know your real fuckin' name; you've refused me that consideration from the very start.

"Billy, please, it's too long a story. There are so many things about me that I can't even begin to explain. Besides, you wouldn't believe me anyway, so let's just drop it! Okay?"

"Well, at least tell me your name," he snapped.

"It's Christopher Sandman Lance."

"So what's the big deal? Why have you kept it a secret?"

"Because I've been sworn to secrecy; besides, you'd think I was fucking nuts; lying, or just plain full of bullshit!"

"Sandman, you look different? You look younger. You even talk and behave smarter then before. You're wearing expensive clothes, you drive a remarkable car; I just don't understand it. If I hadn't known you before, I'd swear you were an impostor. Please, Sandman, tell me-what's happened to you?"

"Billy, I told you before, I can't give you any answers, damn it. Okay. Here; chew on this little riddle for awhile: I was alive in the past, and now in the present, but for years to come I'll be living in the future."

"You're talking nonsense, Sandman, or should I call you Christopher?"

"Call me Sandman when no ones around, Billy. Otherwise, call me Christopher."

"How about I just call you liar!" Billy replied exasperated.

"Billy, I'm through talking. All that you need to know is this: I'm an honest man who has become wealthy, and I still want to be your friend. Now, if you'll please excuse me, I have a date with Darlene. The keys to the liquor cabinet are on the table, and you're going to need to eat. If you should desire anything, the tab is on me. I want you to stay here for a while. I'll secure your possessions and I promise; I'll lock-up your old store. I'll see you tomorrow," I said and closed the door while Billy was starting a new sentence.

The time was seventy-thirty when I drove into Darlene's driveway. I climbed out of the car and pushed the doorbell.

The door opened and there she stood, looking her very best, and with a sweet voice she called out, "You came, Christopher. You kept your promise."

"Yes, it's me, Darlene."

"My God, you look good!" she exclaimed.

Just after she said that, I obviously noticed her looking at my car, then back at me. Here we go again, I thought to myself.

The one thing that the Society hadn't warned me of, and which I hadn't gave any thought to, were the explanations to my new appearance, my car and money. This was a new experience for me-and I dreaded the questions.

"Is that *your* car?"

"Yes, Darlene, that's my car. And before you say anymore, it was given to me as a present."

She stood in the doorway for a few seconds looking, and then turned her attention back to me. Her eyes were showing bewilderment then suddenly she said, "Oh, I'm sorry, Christopher . . . please, come in."

I smelt the fragrance of food, and the aroma was good. I noticed that she had set the dining room table with an expensive tablecloth, and flowers for a centerpiece, including elegant plates, glasses and silverware.

She was wearing a white, satin evening gown, which fit her form exquisitely. It brought out her beautiful figure, which also showed her lovely curves. She looked like a living goddess. My initial thoughts were: why am I here, and why does she care for me so much; I'm not good enough for her. Before I could get the thoughts out of my head, she embraced me with her arms, her body and lips. I was in heaven. After our kiss, we stood back, holding hands and exchanged looks.

"My God, Darlene, you look so beautiful; damn baby, I sure did miss you."

"I was so worried about yesterday, Christopher; I thought I'd never see you again."

"I was behaving stupid. Let's not talk about yesterday, Darlene, okay?"

"Okay," she said with a smile, then squeezed my hands and slowly let go.

"I can't tell which smells better, you or the food. What are you cooking?"

"New England Pot Roast-have you had it before?"

"I don't know, but from that delicious smell, I just know it's going to taste great."

"I hope you'll like it, but it won't be done for another hour or so-I got a late start," she said, and then added, "Can I get you a drink or something?"

"Sure, what do you have available?"

"Let's see . . . I have: coffee, tea, soda pop, brandy, sherry, wine, beer or scotch."

"That's quite a list; I said politely, and then asked her, "What are you having?"

"I'm having some sherry and *you babe*," her voice said softly and sensuously as she appeared from around the corner gleaming. "What would you like?"

"How about some scotch on the rocks?"

"You got it. Just give me a moment."

I was sitting on the couch, next to a burning fireplace and waiting for my drink, when she approached me in a most provocative way, and then sitting down close beside me while handing me my drink. I took a slip and sat my drink on an end table, blindly, while staring into her sexy green eyes. She also sat her drink on the end table, leaning across me and purposefully stimulating me with her fresh hair on my face, while also enticing my senses with her perfume and a well place hand.

"Darlene," I said quite frankly, "I think you're trying to seduce me."

"Now what ever gave you that impression," she said sitting back up with her hand still between my legs, and smiling devilishly only in a way that a deep and caring lover would or could do.

I turned my body around more closely, and then hugged her tightly and kissed her like the French people do. And for the first time in my life, I felt what a *first-love* really feels like. I slowly stood up and excused myself to the bathroom. In there, I took my sport coat off, and more importantly, my maser-gun and holster. I hung them over a hook with my coat concealing my weapon, and returned to the couch. Darlene was lying upright, with her stocking feet crossed along a portion-length of the cushions. I sat beside her feet and began stroking her legs softly while slowly moving toward her thighs. I turned my head to look at her face, and she had her head laid back, eyes closed and smiling blissfully.

"You know what I want to do?" she said after a minute or two.

"Tell me, Darlene . . . please tell me; what do you want to do," I responded with a rapid heartbeat.

She pushed my hands away and gently stood up saying, "I want to *fuck* you."

Hearing those naughty little words, I was compelled to also stand up; showing my vertical erection inside my pants and then said, "Let's go into the bedroom."

"No, Christopher, right here; here on my soft fluffy rug next to the fireplace!"

We made love like a man and a woman should, with no uneasiness nor embarrassment, holding back nothing. The sensations were ten-fold since the first time we made love-and before my mutation. We shared an orgasm together, and she had one more followed by one last ejaculation, I wouldn't hold back. Our lovemaking was the best experience I ever had, or have ever dreamed. I enjoyed the scent of our love making, and she said she did too. We spent some added minutes in our juices, hugging and kissing. After some minutes of sex after-play we got dressed again, and with no concern of bathing but instead simply washing our hands and faces. We then laid back on the living room couch committing our love for each other in words that were never spoken before. Finally, we sat up, occasionally touching each other, while talking and finishing our drinks before dinner.

"What's that buzzing sound?" I asked.

"I don't know," she listened quietly, and then replied, "I don't hear anything."

"Oh! Just a minute." she said while getting up and walking toward the kitchen.

"It's the timer-clock; the food is done; how did you hear from out there?" she asked and continued, "I can only hear it close to the kitchen, but not from way over there."

"Hum . . ." I mumbled out loud, "maybe it came from an echo," I said acting a vague bewilderment, and then replied, "Oh well . . . it's not important."

"The dinner is done-are you still hungry-I hope?" she asked with her head cocked to the side with a questioning smile, and then shrugging her shoulders.

I would have been a real bastard to simply get up and go-but that stupid and ugly thought would never apply to my sweetheart, Darlene. And I was indeed, hungry.

"Of course I am, you've worked up my appetite, I wouldn't miss this dinner for the world-besides, my love, I really am hungry."

We ate and talked, and I'll have to say, the food was great. When we finished eating, we went back to the fireplace and had another drink. It was going on ten-thirty and I knew Darlene had to work tomorrow-I told her I had to be going soon. Although she didn't know it, I had told Billy I'd lock-up his store today, and I didn't want to get there too late.

"Darlene, I'm going to have to leave shortly. I told a good friend I'd do him a favor, and I didn't want to do it in the wee-hours. Besides, I know you have to work tomorrow."

"Christopher . . . before you go, can I ask you a personal question?"

"Absolutely; ask away," I said inconsequentially, and then was astounded by her question.

"There seems to be a subtle change in you-what is it? Something has happen to you because I can see the changes-physically and intellectually. Tell me what has happened, Christopher?"

"You're right, Darlene, something did happen to me, physically and mentally. It happened on that business trip. But I also returned with a great sum of money, a new car and a nice home, and a commitment to write some books. But I have to be leaving now; can we talk about this later?"

"Hold it. What kind of books?" she asked me seriously and showing a genuine interest.

"Well, I found science fiction to be a good genre, so I wrote a science fiction book."

"Have you had it published?" she said with excitement, and then added, "Are you still writing now?"

"Not at the moment, just the one manuscript. As a matter of fact, I'd like to bring you a copy the next time I see you-if that's okay?"

"Of course it's okay. I'd love to read your manuscript, what do you plan to write next?"

"Well, as I just said, I haven't started yet, but soon I'll start writing some volumes on modern science as it is today, and how it will be in the future."

"You're switching your genre from science *fiction* to *fact*? That sounds very difficult. You have to know your sciences and math, and a degree or two would be helpful," she said with a short pause, and then continued, "But I want to be honest and sincere with you, Christopher, and with all due respect, you do appear more intelligent then the first day I met you-oh, I'm sorry for saying that," she said apologetically.

"No need to apologize, Darlene, but now I have to go. We'll talk about it later, alright?"

"Are we going to talk about the book, or you?"

"We'll talk about me, but I'm worried I'll lose you if I tell you the truth."

"Don't be so ridiculous, Christopher, I've grown too much in love with you."

"Say what you will, but I'll still be praying you won't leave me," I said as I walked to the bathroom to get my coat.

After I thanked Darlene for the best night of my life, I kissed her goodnight, and then started for Billy's store to lock-up.

God only knows what's been stolen since he's been gone today. For the first time in quite awhile, I'd be back on the streets after midnight. My stomach started to feel a little queasy, and I attributed it to being out at night again, with a new kind of fright after being away for so long. I was no longer a bum, a thief or even a murderer. I resolved myself to a new life, and that felt really good. However, I could still feel that little queasiness; something didn't feel right. Maybe it's because of my alertness and my improved senses. I wasn't sure. But, to give an example, I could see as well in the dark as I could see in the day, and that's never happened before.

I parked my car in a well-lit parking lot, about five blocks from Billy's store, and then proceeded to walk the rest of the distance. My senses were on high alert. From a block away, I could see a light inside the store, but not upstairs where Billy and Betty lived. Slowing my pace, I focused on the inside of the store. I could see people inside. They were clear and sharp outlines when I squinted my eyes, just as if I were using my binoculars. As I approached the store from across the street, I stopped for a few minutes to think: how should I handle this situation?

I thought about simply calling the police, and let them chase whoever was inside away. But I didn't really want to get involved with the police-too much information exchanging. I thought it might be better if I'd just chase them away myself-lock the store and chalk it up as charity.

However, if something went wrong, it could make the news and screw up our new store deal. Just then, a thought occurred to me-a new scheme. It was a very simple idea: I'll spook them out. This, I couldn't have done with my old .38, but it might be possible with my maser gun. I was told that with the power set to the non-kill position, it feels like an awesome bee-sting.

Pulling my maser from its holster, I positioned it to non-kill. I was beginning to think that this might be more fun then scary. I rounded the block to approach the store from the rear where there were some bushes I could hide behind. I also positioned my location to the side of the building where the steps led up to Billy's apartment. I saw two people sitting on the steps, sharing a crack pipe and drinking liquor. Although I'd rather the steps were empty, it did pose a good opportunity to test my scheme. I took aim and shot. Not a sound until I heard one of the guys on the steps scream.

"Yhoooooo! Son of a bitch! Something stung me right on the ass!" he yelled as he pulled down his pants and swung his ass in the face of his partner and whining, "do you see anything?"

I took aim for the other guy.

"Ouch! What the hell was that! Damn it; that fuckin' hurt," the other guy said while turning around to look back. I knew they couldn't see me. One more should do it, and so I took aim at the first guy again.

"Ouch!—On my leg, Oh! It hit me on my leg. Jesus Christ, it burns! I'm getting the hell out of here-did you hear anything?"

"No," the other guy replied quickly and panicky.

"Fuck this shit . . . something's not right. I'm gettin' outta' here!" he said turning around, only to see his buddy already running away.

I stood up and started to giggle quietly to myself. I felt like a little kid playing pranks. Now that I had my entrance clear, I could make it up to Billy's apartment. When I reached the landing, I discovered that the doorknob was locked. I twisted it hard until it broke off in my hand, and the door slowly swung open. I stepped in like a ground soldier in the Special Forces. All the lights were off, but that didn't bother me-my eyes adjusted quickly. I quietly walked around and surveyed the rooms to make sure that nobody was up here with me. The apartment was safe.

I could hear the people downstairs in the store, laughing and talking like it was a New Years' eve party. I listened very carefully, and was able to distinguish between eight to ten different voices. I averaged nine people. I opened a door that led down to the steps and where the people were. As I causally stepped down the landing, I had remembered that Billy had put up that bowl-curved mirror at the bottom of the steps-and for obvious reasons. This shining gift would give me an added little edge to spook these bums out of the store.

I sat down on the stoop in the dark knowing that I could see them but they couldn't see me. Easy pickings. I even thought that this might be amusing. I took aim in the mirror and focused my maser on some guy's thigh. This was too easy. I fired my gun knowing that my *bullets*-those sightless, soundless streams of photons, would simply bounce off the mirror and onto my enemies' reflection, but only if I made perfect calculated aims.

"Ouch—shit! Goddamn it! Who did that-feels like a fuckin' snake bite! Who's the bastards playing tricks?"

I heard the talking and laughing die down, and from the mirror I could see some of their expressions. I was starting to smile. They hadn't a clue to what was going on, and that made me feel even more zealous.

"Who stung me!" some guy demanded.

"I don't know what you mean, but it wasn't me man," I heard some guy say.

"I didn't hear anything. It must've been a rat biting your ass.

"Chill!" someone else shouted, "Don't get something started," and then added, we're trying to have a party here damn it!"

I took aim again. This reminded me of billiards-bank shots off a mirror instead of bank shots off pool cushions. I shot again.

"Son of a bitch! Right on my balls! Oh . . . shit it hurts, he said bending over. Who's fuckin' around?" the bum yelled with a furious determination of a revenge killing.

After those words, I started shooting them one at a time quickly-mostly on their ass or crotches until they were all jumping around like stupid string puppets operated by monkeys.

I couldn't help but start laughing bizarrely when they all began scattering out the front door. And I'm sure that my laughing also added to the strangeness of the whole scene. It was only a matter of a few short minutes before the entire store was clear. I checked around to make sure no one was there, and then locked the front door. I then switched the lights off and walked upstairs to fix the door that I had broken. I stood on the outside landing, jamming the apartment door closed securely with a strong piece of steel. I walked down the steps laughing, slapping my hands clean, and thinking: that was much more fun than calling the police.

I walked in my new home and threw my keypad on a table. It took me a while to settle down. I had a drink and tried not to think about it. God, it felt good to be back home again, in my living room having a nightcap, with plans to read a book I found on chemistry, and then later going to bed. And I sure didn't miss that old apartment and the stinking neighborhood. I felt safe here, but in an unusual way-like *Nosferatu* hiding under the safety of his closed coffin. The change in my lifestyle was sudden, but good. I could sleep here, and so I did.

I went to bed late, but I woke up rather early. It was at eight in the morning when I had my first cup of coffee. Feeling refreshed and ready for the new day, I decided to make my bed, clean up and get my morning routine out of the way. I wanted to see Billy, but before going, I spent a little time exploring my home. After only two nights in my new home, it already knew my voiceprint and was beginning to learn and respond to simple commands I gave it. It grew on me very quickly.

Initially, the house seemed very large and elaborate. The wide, red-carpeted stairs was a real eye catcher. Some of the rooms were empty while others had furnishings but all had built-in technologies; lights came on automatically and the house would open doors and windows and a few other things simply by my asking. Each room had its own personality, and in the time of my explorations it appeared to have grown smaller as I became more familiar with the place.

I realized I was still responsible for my own cleaning, washing laundry and other household chores-but that was trivial to me. Hell, if I wanted to, I could always hire a maid. I ventured back to my living room, when suddenly the phone rang. With anticipation I picked up the receiver and immediately said, "Hello, Bradly."

"How did you know it was me? I didn't even give you a chance to . . ."

"Just a wild guess. Only the Society knows my phone number, and it wasn't too difficult to figure it would be you-my mentor. How man I help you, Bradly?" I asked him cordially while remembering how nice he was to me.

"I've talked to Dan and David this morning and they have all the necessary papers complete and ready for both, yours and Billy's signatures. They've been devoting all their time and attention exclusively to you. I'm sure you won't mind, Christopher, but I took the liberty to set another meeting for you and Billy at twelve today-same hotel, same conference room. After today, you and Billy may do with the building as you please. Can you make it there by noon?" Bradly asked in such a way as to expect only one answer.

"Yes sir; I'm on my way, Bradly," I said still talking as the phone was on its way to the receiver. I paused for a moment and then immediately dialed Billy's number.

The phone began ringing, and kept ringing and continued ringing until I could stand it no longer. To hell with it, I told myself, I'd just drive over there. If we don't make this meeting we could lose the whole deal . . . and that I was determined not to do. I reached his hotel door and began pounding.

"Billy, wake up!" I shouted and then pounded some more while waking the neighbors whom I just ignored and continued to pound while telling the nearby, angry residents to go back in their rooms, "this will be over with very soon." And with all my might, I forced the steel door open breaking the lock and jam using the new muscles in my legs and shoulder. I walked in and found Billy on the floor.

At first I thought he was dead, until I smelled the alcohol. He was confused for a few minutes, and then started to regain his surroundings and me. Needless to say, Billy was suffering with a terrible hangover. I blamed myself. I'm sure he got drunk last night because of me, when I shut the door on him so disrespectfully.

"Come on, Billy, get up; we have a surprise meeting today at noon to seal the deal," I told him while slapping both sides of his face lightly with the palm of my hand.

"Meeting? Quit hitting me!" he said with a harsh and haggard voice.

"It's ten-thirty right now, Billy, and we haven't much time. We'll have to get you cleaned and sober by noon. That gives us an hour and a half. Do you want that new store, Billy?—Because if you do, I'd suggest that you start making a concentrated effort at getting your head on straight and start for the bathroom. I'll help you in the tub and turn the shower on. Come on, Billy; get up! For the last time-do you want that new store?"

"Help me off this fucking floor!" he bellowed, and then said less loudly, "help haul my ass to the toilet."

I could hear the poor old bastard in there splashing water on his face, along with snapping farts, squirting-shit splashing, puking, gagging, coughing and making other nasty sounds that were just as disgusting. Thank God all the bodily-noises quit after about thirty minutes-just in time for hotel-security to crash our little morning-sickness party. There were four men standing inside and outside the broken door.

"What's the problem here, sir?" one of the four said standing inside the door.

"I thought my friend was hurt, or had a heart attack. He didn't answer his phone when I called him earlier, and he relies on a wheel chair. I came here as fast as I could and broke in the door for fear of my friends safety. But now in hindsight, I should have called you guys first; I'm sorry, I just wasn't thinking. I hope you men can appreciate my concern for my friend.

The hotel room is in my name. I can make retribution for the damages later today, and if you should absolutely need me, I'll be in conference room 503. I can assure you, gentlemen, this will all be taken care of before the day is over. Is that okay?" I asked the security men politely, almost begging self-shamefully.

"I suppose that could be arranged, sir," he said unperturbed and showing little interest.

"Thank you," I said, "Now if you wouldn't mind, my friend is in the bathroom getting cleaned up, and he'll need my attention. We have an extremely important meeting today at noon and we're running late, so if you please, sirs, let us continue-and my apologies for all your troubles," I told them, half lying, half truthful.

"Go ahead about your business; the desk clerk will have a bill for your damages. In the meantime, and as soon as you leave, we'll have all your items moved to another room. Check with the desk clerk later," he said quite professionally, and then walked away while the others followed.

I turned my attention back to the bathroom, walked up to the door and said, "Are you okay in there?"

"I hear yah; just give me another few minutes or two," Billy said convincingly.

I waited five minutes, and then said again, "Are you sure you're okay in there?"

"Yes," he answered more clearly, and then asked, "Can yah put me in the tub?"

"On my way," I said approaching the door, and without even thinking, nor expecting, all of the puke, piss and shit on the walls and floor. It was definitely an eye-full but more a nose-full when I first walked in.

"Good Lord, Billy, this is awful . . . look at the place-are you sure you're okay? After this meeting, you're seeing a doctor-I want you checked over," I said while lifting Billy into the tub, and trying very desperately not to get any shit on me.

"Fuck those goddamned doctors!" he yelled at me while I handed him a shaver, toothbrush and cream.

"What do you mean by saying that?" I sniped, "doctors are good."

"Very good at making money from people who can pay it, but not for dumb fuckers like me."

"Call me when you're done," I said blatantly, ending the conversation.

Only after the real estate agent and the attorney left, did we hug and shake hands, while sharing a laugh-like we'd just won a lottery. It felt good, not only for me, but also for Billy. I could somehow feel the freedom he felt, like a man who's just cheated death. I also sensed he felt more full of life. He was now a new person, and he knew it-without any extreme burdens, like something you could only pray for.

I felt that he knew he was now out of a very deteriorating cycle-not to the point of destitution-but to the point of resolution.

CHAPTER FIFTEEN

Later on, about four o'clock, I was rubbing my feet and toes through the deep shag fibers of a silk carpet. I was in my office writing a manuscript on science fiction. Because I hadn't yet enough knowledge of science fact, I decided to write something about science fiction. I read a book about *Super String Theory*, which is hard to understand, but I understood it because it was in layman's terms. It had a few interesting ideas which helped inspired me to write a book. My skills with computers, especially on the keyboard, were improving very rapidly. I was typing over 200 wpm with no mistakes. I spent fifteen hours on my first science fiction manuscript. I titled it: *The Calabi-Yau Universe*. The time spent on the manuscript made me tired. It was seven in the morning when I went to bed.

I woke up with a rush, sitting straight up in my bed. It felt like I was late for something. It was four in the morning and I had slept through an entire day. There was an overwhelming feeling of guilt, and even a twinge of fear. Like I had just missed reveille, or some important meeting. But I hadn't planned on doing anything yesterday; so what the hell-I got over it.

When I made it to the kitchen, I *told* it to make me a cup of coffee-one teaspoon of sugar, two ounces of cream.

I spent about five hours reading and working on my computer, and fifteen minutes to eat. I put all my dirty dishes, scraps and all, into an opening below my sink, and they return back to my cupboards and drawers, all nice and clean-I knew how it worked. All of my groceries are sent to me by making a special telephone call to the Society. I have hundreds of suits and other clothes, which fit me perfectly and tailored for my dimensions by the Society. I wear a new outfit everyday, and I see to it that my worn clothes go to charity.

There were basically three attributes from the operations, which were starting to change my life. I could already feel more strength, and an increase in my senses, but my brain was another thing.

It wasn't developing as fast as my senses, but they were correct at the Society; I was indeed growing more intelligent because of my insatiable interest for knowledge. From the confines of my office and home, I grew more need and desire for science information. The other night, I was surprised how quickly I learned the principles of advanced algebra and the beginning of analytical calculus. As I got deeper into it, I read a technique on how to solve mathematics equations on a computer. I found that numerical programming was much quicker. It proved to solve some of the most complex equations I've yet to work with. I also found all the other sciences just as interesting. Before my operations, my brain could have never begun to understand the principles of science. But now, I could understand them by reading alone. And in my little spare time, that's what I did.

After about ten o'clock or so, I made a bold decision to clean myself up. Afterward, I was sitting in the living room wondering whom I should see first: Billy or Darlene. At this point in my change, they were the most, well . . . almost, the two most important people in my life. I decided to visit Billy first, because I knew Darlene would likely be at work.

I hopped in my car, drove it out of the garage and down my driveway until I came to my stronghold-my gate.

After a push on my keypad, I was on the road and making my way to the *Emerald Hotel*, and Billy's room. Instead of pushing in his door, I had the desk clerk phone his room.

"Yeah?" I heard Billy's voice from the clerk's receiver.

"I have a gentleman here to see you, Mr. Richers. His name is, Mr. Christopher Lance; may I send him up?"

"Yes, please," I heard Billy answer.

The first thing I saw, as I walked into the new room they put him in, was Billy sitting at the table, in deep thought and looking concerned. He apparently had been crying and I could also tell that he had sat up all night drinking. He was still wearing his suit.

"Billy, what the hell's wrong with you?" I asked confused, but compassionately while I put my arm around his shoulder.

"A lot of things, Sandman, I lost my wife and I miss her badly. I still cry when I'm alone, because I keep thinking of her . . . I'm lonely, and I'm scared. I'm beginning to believe that nothing really matters-no sense of purpose, and no direction to follow. You weren't around when they killed Betty. You don't feel the horror and pain, or that black empty sadness I feel every day. Can you show me how to grieve . . . Sandman? Can you help me make sense of it all, will you talk with me and . . ."

"We'll talk about it, Billy, but not now, give yourself another few weeks to grieve and for the sadness to get out of . . ."

Billy interrupted, "And I don't want anything to happen to you, and that makes me feel selfish and ashamed. If something should happen to you, I'm afraid of what's going to happen to me. I'd be right back where you found me. The idea of owning a new store felt so good yesterday, but now, I just don't know-I'm not sure if I can do this thing on my own. I don't have the money or the expertise. I don't even have a home. I appreciate all that you've done; don't get me wrong, but where do I go from here, Sandman?"

"Maybe this will brighten your spirits up, Billy; you'll by my guest here at the Emerald Hotel for a few more days. What did ya think I was gonna do; put you out on the streets?" I laughed while patting his back, "I understand that you're still grieving, and that's okay, but don't worry about money or tomorrow, or the days to follow. As a matter of fact, I'm buying you a home near your store-I wanted to surprise you. You'll be able to move in in a couple of days after I'm through having it refurbished and furnished. Oh, and I almost forgot . . . I've ordered a special van with a lift for the front seat and a motorized wheel chair.

You'll be able to drive, using basically only the steering wheel-the gas and brakes are attached."

"But the money? Where will that come from?"

"Where do you think it's been coming from? I've got you covered . . . so don't worry about it. Now, let's start with the fun stuff-have you decided on a name for your new, and first ever super market store?"

"Well . . . not really. It doesn't seem real yet-more like a dream. Did we really buy a store?"

"Hell yes! Have you taken a look at the floor plans I left you?"

"Well, no. There was nothing . . ."

"Well nothing my ass; let's get down to business, my friend. Let's think of a name for you new store; you go first," I told Billy, and he began to smile.

"How about, 'Billy's Super Market'," he replied.

"No, to mundane, makes it sound like a small store. We need something that sounds bigger and easier to remember. How about this one, 'Billance Grocery Store', your name with mine? Bill-Lance."

"Hey, that sounds better, what about; just the word, 'Billance'," Billy said.

"Yes! Yes . . . I like that," I exclaimed, "one word, easy to remember, easy to market and advertise. Within a month, you'll be doing business like you've never seen before. Okay, now let's look at the floor plans."

We discussed the square footage and dimensions, finding areas for what should go where and also talking about what kind of items he would stock. The list he recited to me seemed to go on, and on, and on. Finally, we reached a point where we were deciding what kind of people we'd need to run the store: managers, stockers, sales, meat cutters, cashiers, baggers, janitors and even security. We covered it all. I told Billy to set a date and I'd run a full cover ad in the newspaper for hiring. He sat a date for next week, and I asked him if I could be with him when he makes the decisions to hire. Not surprisingly, he said, "yes"-but actually, I think I heard the word: "please!"

Before I left Billy, he was in a much healthier mood. I think he was finally realizing that this was for real. While I was still in his room, I called the company that sold me his deluxe model, motorized wheel chair, including a lift seat for help at getting in and out of his van. They told me it would be at Billy's room no later then five.

The time had come for me to leave and as I was closing the door, I peeked around and said to Billy, "Buy some new clothes tomorrow, valet will have your new van ready, and for heaven's sake, wash yourself up and get something to eat, and then get some sleep." He smiled when I left.

On my way to Darlene's house, I realized that I had spent the entire afternoon with Billy and enjoyed it. Minute by minute, hour by hour, I could see him becoming more excited. Strangely, there were a few times near the end of my visit when he was talking, that my hearing just kind of faded away into silence. All I could see were his eyes blinking and his lips moving; like tunnel vision, or something like it. That's when my mind would drift away. Although I could see Billy's face, I was in a completely different place. This was by no means, a blatant attempt to shut him out. His words just seemed unimportant compared to what I was trying to think about. Suddenly, my lapse into this state of unease would stop immediately, and I was back in Billy's world again. This strange experience occurred two, or maybe three times while I was with Billy. There was something in my subconscious that was trying to make its way out, but never did.

On my way to Darlene's place, it hit me. I now understood what those lapses of uneasiness were. I forced it out into consciousness. Guilt. I was feeling guilt. I was guilty of being untruthful, deceitful, spending huge amounts of money unceasingly and keeping my secrets from Darlene.

It was a strange dichotomy; having my life completely turned around in such a positive way, and yet, having a dark secret to one slice of my life was like a weight over my shoulder that's suppose to remain invisible. It would be impossible. I felt cleansed and dirty at the same time, but now I wanted the dirt to go away. How long could I keep my hidden past and untold life concealed? Hell, even I could see the physical difference in me, let alone, Darlene. My body was more muscular, appearing as if I had been lifting weights. And my beard wasn't as thick before; I indeed looked younger-maybe two years. I couldn't keep my change a secret anymore. I didn't know what I'd say to Darlene, but the guilt was really getting to me. I was afraid of losing her, but yet, from nowhere, my biggest unease was the Society.

When I arrived at Darlene's, she was sitting on the porch smiling, as if she knew that I would be arriving at this very moment, and waving me a hello with her hand by closing and opening her fingers and palm together. She stood up, wearing blue jeans with a short T-shirt, as she approached the car the engine stopped.

"Christopher," she said as I opened the door, "take me for a ride in your car."

"Well, okay, climb in and fasten your belt."

I started the engine, shift to neutral and pumped the gas pedal, making the motor come alive again-it roared like an angry beast. Darlene showed overawe at the sound.

I've never thundered that mighty engine in front of her before. I slowly backed out of her driveway and headed west down Steward Street. The highway was quite busy, so I drove at normal and lawful speeds. When I reached the mountain roads, and only on a straight highway, did I crank through the gears and drove the engine to a scream. At about 170 mph, Darlene started to yell, "Slow down Christopher! Please! Slow down-you're scaring me."

"Okay, okay, I'm slowing down. There. You see-120 mph and still slowing. I didn't mean to frighten you, I thought you wanted a thrill ride," I said innocently.

"You're risking death!" she replied agitatedly. That was the first time I've ever seen Darlene upset with me. I felt embarrassed by my action. Darlene must have seen the dissatisfied look on my face; feeling stupid and childish.

"Christopher, it wasn't just the speed, but also your face. Your expressions scared me. You didn't look like you. You looked mad and angry going so fast; that also scared me. I thought you were going to lose control and kill us both-I thought we were going to crash-sorry I yelled so loud."

"That's okay, I deserved it. I was behaving like a fool; like a young, spoiled, rich kid with a new Corvette his parents just bought him. I'm sorry. I won't do it again."

"Thank you," she said, "you could've killed us, and then I'd never speak to you again," she said with a tiny smile. She broke the tension.

"Any special place you'd like to go?" I asked with a note of shame, not sure what else to say.

"Show me your new home," she replied abruptly.

Her face was sincere and honest, but with a degree of uncertainty, as if she wasn't sure if I had told her the truth. I decided to let the truth speak for itself because I knew this day would come soon. I made a U-turn in the middle of the road and started for home driving carefully and slowly. When I pulled down Salisbury Lane and drove by the first few homes, Darlene's eyes started to grow bigger and bigger, with each new home we passed, which also grew bigger and bigger. At last, I pulled onto my driveway and stopped at the gate.

"Where are we, Christopher? We'd better leave before we get in trouble. Look at that huge gate, and I don't even see a house. This is some type of security gate. Take me back home. You don't live here-now let's just . . ."

Before Darlene finished her sentence, I had pushed the button, which she didn't see, that opened my gate.

"Oh shit! Now we're in trouble. Hurry up, Christopher, let's leave-the gates are opening. Someone will be here soon!"

Instead of putting the car in reverse, I put it in first and drove through the gate. Darlene quit talking immediately, and just looked at me as if I were insane.

"What the hell are you doing?" she whispered loudly.

"Relax," I said calmly, "this *is* my home, Darlene; trust me and you'll see."

Darlene sat quietly in the car as I drove up my driveway, looking at all the trees and bushes along the manicured lawn, and all the spring flowers that were starting to bloom. As we continued up the slope of my drive, she saw, as I did, the third floor roof of my castle home rise from the horizon. Her eyes grew bigger, as my house started towering into view. I stopped the car in the oval turn-around in front of my home.

I got out of the car first, and then opened Darlene's door to escort her from the front seat. In the next few seconds, we found ourselves standing near the front door. Darlene looked left, then right, then up with a hand shielding the sunlight on her face.

"Oh-my-God!" she said as she pivoted 360 degrees until she was facing the front door again.

"Would you like to come in?" I asked casually.

"Oh, my-yes-I would," she said as I smiled and held her hand.

When we went in, I showed her everything. I showed her my living room, kitchen and dining room, my bedroom and bath, my study, the guest rooms and even my office. Like me, she was awed and even a little insecure by the magnitude and complexity of this place. As we toured my house, she would say simple things like: "wow, oh that's so beautiful, I love those colors, such a gorgeous rug, goodness; it's so big," and other such things. We finally ended the tour back in the living room, where I made us a drink.

"Please, Darlene, have a seat. What can I get you to drink?"

"I'll have what you're having," she simply said.

"Are you sure? I'm having a glass of scotch."

"Then that's what I want," she said sprightly.

I handed her a glass of scotch and sat in a chair next to her. She took a rather oversized gulp and stared at me in a strange and peculiar way. We sat in our chairs quietly for a moment or two, and I could tell by her expressions that she was trying very hard to take it all in very quickly.

"So, my sweetheart, what do you think?"

"I'm lost for words, Christopher; I don't know where to begin, or what to say. To be quite frank, I'm completely confused. I just don't know what to think?"

"Do you still love me after seeing all this?"

"Yes, I do, but my God, Christopher, where in the world did you get so much money? Please don't tell me it was an act of evil, while you were gone away."

"Oh, no! Nothing evil has taken place. As a matter of fact, it was quite the opposite. Something beyond stupendous has happen to me, and I'm not sure if I can even begin to explain. Actually, I'm not sure if you'd even believe me," I said honestly.

"Try me," she said, followed with a slip of her drink.

"I'd like to. I want to, but . . . I'm not even sure if I can, or should," I replied perplexed.

"Then you are hiding something from me. People just don't get this rich so quickly. Now I get it-you are up to no good! And if you aren't, what is so bad that you can't tell me what's going on. Will somebody hurt you; kill you perhaps? I demand that you take me back home-unless, of course, you plan to kill me?" she expounded angrily, and without rationality.

"Darlene, you're talking nonsense."

"Why should I believe anything you say? This house could belong to somebody else. Maybe even your car. This is too much for me; it doesn't make sense and you're starting to scare me. You could be some kind of psycho-a serial killer."

"Oh, for heaven's sake, Darlene. You're blowing this all out of proportion. Psycho? Kill you? Don't be so ridiculous. Jesus Christ, I love you! You insult my love and intelligence. If you wanna go home, then let's go. I don't need this insecure bullshit. You sound like you wanna be done with me? Okay. Fuck it! I'm through with you too. And believe what I'm telling you-once I make up my mind-that's it! You're outta' here, including my life! I'm so goddamn mad I can't even drive. I'll call you a cab," I kept ranting angrily.

As I started to walk toward the phone, I felt something bump off the back of my head. It made me stop and stand still. I saw bits and pieces of shattered glass and splashes of scotch fly pass me. I slowly turned around and saw Darlene standing with her hands over her mouth. I descended quickly into an abyss of utter sadness. It hurt my feelings more than my head. I tried to verbalize the word why, but couldn't.

"Oh my God, Christopher! Oh my God! What have I done? I didn't mean to hit you-honest! Please don't hit me. I'm sorry. Please don't hurt me; don't beat me-I'm so sorry-I didn't mean to hit you. I didn't mean to hit you. Let me see if you're bleeding and I'll help you.

Oh dear God, oh dear God, what have I done?" she said squealing and crying, while sinking to the floor, until she was curled up in a ball with her arms and hands over her head; waiting for her beating.

My heart went out to her. I felt so sorry for her. This was how her old, dirty-dead husband, Vern, has trained her. It was a conditioned response. It was her action, not her emotion that troubled me. I walked over and bent down to pick her up gently. She stiffened her entire body, especially her arms over her head, which I supposed was her defense in preparing for a beating.

"Relax, sweetheart; relax your body. Put your arms around my neck and I'll carry you to the sofa."

She slowly moved her arms away and around my neck, and then looked at me with tears and mascara running down from her eyes to her cheeks. Within the second or two that I held her, she blinked and jerked her head slightly a few times, and I couldn't understand why. Then it suddenly occurred to me: she was waiting for my fist to hit her face. I replaced an expected fist with an unexpected kiss and pulled her up gently, laying her softly on my sofa. I wiped the tears from her face and hugged her tightly.

"Oh, God. I'm so, so terribly sorry I threw that glass at you, Christopher.

Please turn around so I can have a look, you'll probably need stitches, and I'm sure you're bleeding badly," she whispered softly with trembling fingers touching the back of my head.

"Don't worry, Darlene, I'm okay."

"Your head isn't bleeding; it's just the wetness from the scotch. Oh, thank God you're okay. But maybe we should have it X-rayed? I might have hurt you inside or something."

"It's okay, I promise, I won't even get a bump on my head. Why did you throw that glass at me?"

"I don't know; an old habit I suppose. It's the way Vern and me use to fight-yelling and screaming and throwing things at each other, but I promise you, Christopher, and I'll even swear on the Bible, I'll never, ever show that kind of stupidity again," she said softly while sitting up next to me with her arms still around my neck and leaning her head.

"It's was my entire fault, Darlene," I said, "I haven't been myself today. And you've been right all along. I haven't been honest with you, and I feel ashamed for my actions and words. I'm afraid that if I tell you the truth about me, you'll leave."

"Okay, Christopher, okay. I won't ask you any questions, but how will I ever believe anything you say to me in the future? How will I ever know when you're telling me the truth?"

"I understand; I know what you're saying. You're right, of course, but I'm afraid of the truth as much as my secrets. I'm afraid that if I tell you the truth about me and how I obtained my money, you'd still think I was lying."

"Try me."

"Ah . . . I need another drink," I said.

"You do that. In the mean time, may I use your bathroom to clean my face?"

"Of course. Down the hall and to your right."

"I remember; just give me a few minutes."

"Would you like me to make you another drink while you freshen up?" I asked.

"Do you have any sherry?"

"Let me see . . . ah yes, I have some amontillado; would you care for a glass of that?"

"Perfect. I haven't had that in a long time and it's one of my favorites."

When Darlene left for the bathroom, I quickly gulped down a half of pint of scotch right from the bottle. She must have been in there for over ten minutes. I made her drink, and finished my pint. Nothing. Not even the slightest buzz. Alcohol did nothing for me-I was immune to tiny doses. I could only drink it for the flavor and taste, but not for its effect. It was like drinking soda pop, with the exception of my breath. I fixed myself another bogus drink, sat down on the sofa and waited for Darlene. I needed something to distract me, maybe some music.

"House?" I commanded, "Give me some soft music. Lower. Lower. Dim the lights for me. That's fine."

"Who were you talking to," Darlene said as she entered the room again.

"Just the house."

"Oh, yeah. I forgot about that," she replied," Do I look okay?"

"You look very nice, as usual," I answered.

"Oh, we have music. That's nice. Is this my drink?"

"I made it just for you."

"That's sweet," she said, then added, "Well?" then silence.

"Well, what?" I asked after a few moments.

"Are you going to tell me how you made so much money, including your car and home?"

"Ah, yeah, but . . ." I said as I stood up, "Can you keep a secret?"

"For you? Of course I can," Darlene said with a straight face while crossing her arms.

I looked at her puzzled.

"What?"

"Darlene; I'm being very serious now. Don't make this some kind of game-believe me, it's not. My life could be in jeopardy if my secrets ever got out."

"All right, Christopher, I'll listen seriously, and I promise to tell no one."

"Thank you," I said, crossing my legs tightly while holding back the pint of scotch I guzzled, "But before I begin, I need to use the bathroom, and quickly."

"Go . . . hurry, do your thing. I'll wait."

I trotted to the bathroom just in time. "Ah . . ." I said to myself, "Too much scotch too fast." As I was looking at myself in the mirror, an idea suddenly came to me. After I was through peeing, I took all my cloths off. This might help.

I was bigger then the last time she saw me naked: my biceps, triceps, deltoids, gluteus, quadriceps, and all the other muscles in my body looked sculptured, more like a weight lifter and it was very attractive. However, she wouldn't have noticed with my clothes on, so I left the bathroom naked.

"Okay, now look at me," I said to Darlene, standing poised with my hands on my waist.

"God . . . you're beautiful!"

Her eyes were all over me, and then stopped to focus in one direction, which I realized, at first unknowingly, that she was studying my privates.

"Not there, but everywhere-look at my body," I said with a twinge of momentary self-discomfort.

"I did! You've obviously been exercising. I *can* see the changes and it's very becoming; now, put some cloths on-you're either enticing me or embarrassing me and I don't know which," she said with a sexy grin and then with a smile.

"Darlene, take a good look at my face," I said bending down to kiss her, and then backing away a few inches.

"Oh, you've also been exercising your face, if that's possible. It looks nice, kind of younger, or smoother. Oh . . . I get it-you've had electrolysis; your beard isn't as heavy?"

"I haven't been exercising! I've never had electrolysis, and I haven't been lifting weights! And yet, even at this very moment, I'm still changing. Do you see the difference?"

"Yes, I do! Have you been taking steroids?" she asked with concern in her voice.

"No. Nothing-that's my point, it's one of my secrets. Here, this may help to convince you. Please, Darlene, scoot over to the middle of the sofa."

She did as I asked. "Okay. Now what?"

"Just sit still and don't move or panic."

I walked around behind the sofa, directly behind Darlene. I squatted down with my back straight, and extended my arms under the sofa like a fork truck. It wasn't much trouble, or even strenuous, but I slowly lifted the sofa over my head while Darlene panicked.

"Christopher! Christopher! What are you doing? Oh, good God-this is impossible! Don't let me fall-okay. Lord, this is fantastic! Okay, okay, Christopher, please put me down now. How did you do that?"

"With my body," I said as I let her down gently.

Within a split second, after the sofa was down, she was on all fours looking underneath the sofa. She examined it closely.

"I give up. What's the trick-where's the mechanism?" she asked in bewilderment.

"It was me; no trick. Have you ever seen anyone do that before?"

"No. But I've seen men lift heavy weights on TV."

"All right, okay, one more time, but without you-you're to light. Do you think you could pick up that chair?"

"Oh, a test of my strength; what are you going to prove with that?"

"Nothing. I just want you to feel how heavy it is; try and guess its weight."

She approached the chair and tried to lift it, but couldn't. "Okay, smarty-bare-ass, now what does that prove? I'm weak?"

"No. Well, yes actually, but I think you're missing my point. I want you to tell me what you think it weighs?"

"If I have to guess, I'd say it weighs about 125, maybe 150 pounds max. I agree that it's heavy."

"How about the sofa; try and lift it, or even move it, and tell me what you think?"

"You know I can't lift that thing."

"Then move it; push it, and try to guess its weight," I stressed again.

After a grunt or two, she managed to push it a few inches. "Okay, this bastard is heavy," she said looking at me curiously.

"What's your guess?" I asked.

"At least, I would suppose 300 pounds or more. But I don't understand. You're not making any sense; what are you trying to prove?"

"Just stand over there and watch," I replied confidently.

There were three chairs that matched each other, and one of them was the one that Darlene tried to lift.

Without effort, I place each chair up side down on the sofa-one in the middle, and two on the ends. All three were snug and balanced just right to give me a good center of mass. As I did before, I lifted the sofa above my head with my arms stretched out. Without shaking, and with a strong, tight stance, I looked at Darlene. Her eyes were wide open in disbelief, maybe even in denial.

"How much weight do you think I'm holding," I asked in a non-quivering voice.

"Oh-my-God! Oh-my-God! Put 'em down, Christopher, put 'em down before you hurt yourself!"

"First, you have to guess," I said defiantly.

"I don't know. I can't think. Please, just put them down before they fall on you!"

"First, you have to guess," I said again.

"Okay, okay, let me think," She said hurriedly, "about 1,200 pounds. Now, please put 'em down, Christopher!"

Without moving, I made a mental conversation. "Let me think for a second. That would be about . . . well, let's just say, half a ton. Does that sound right to you?"

"Yes, yes, that's good. Now please, Christopher, put them down! You're scaring the shit out of me!"

"Have I made my point? Do you think I'm different than other men?" I asked, wondering if she believed me.

"Yes! Yes!" she shouted.

I slowly set the sofa back down on the floor and replaced the chairs to their original places. Darlene was still standing in the same spot, with the exception of the expression on her face. She was as white as a ghost. I didn't realize how much I did indeed frighten her.

"Darlene! Are you okay? Here-sit down, I'm sorry if I scared you. I didn't think you would react this way. Please . . . sit down. Here-take a slip of your drink and try and relax. I'll be right back; I'm just going to quickly put my clothes back on."

It wasn't even a minute before I returned, still tucking, buttoning and zipping my pants. I sat down next beside her. Thank God, her complexion was coming back. She might have been in a state of mild shock. But this was something I had to show her. She wouldn't have believed me if I just told her. She wanted the truth about me, but, unbeknownst to her, I had just started.

"Darlene . . . are you okay?"

She slowly turned her head and looked at me. "That was the most awesome thing I think I've ever seen," she said with a face I'd never seen before. "How did you do it?"

"That's part of what I have to explain to you. You wouldn't have believed me if I simply told you, or would you?" I asked.

"Noooo," she slowly said.

"Now, that's what I was trying to tell you earlier. My life, my body, my mind, my money, my home and car; will you believe what I tell you from now on?"

"Yes. Yes, I will-just give me a moment . . . please."

I sat quietly and patently waiting for her to speak again. She kept looking at me. Finally, I broke the silence and the strange stare.

"Do you still love me?" I asked worriedly.

"Yes," she replied quite frankly, "but this is very interesting, I want you to tell me more. You've convinced me, please, continue with your story."

CHAPTER SIXTEEN

"Okay, all right. You told me at your dinner table that you discovered that I wasn't as stupid as you thought. You're a writer, an editor; you have a degree, you're smart; ask me a question, or even a word you'd think I wouldn't know?"

Suddenly, she became much more alive, and then said, "Who was the ninth president of the United States?"

"Too easy, William Harrison, 1841, but he died in office; ask me something more difficult?"

"What is a derivative?"

"In context, please?"

"Chemistry."

"Okay. It's a chemical substance made from another chemical substance-usually organic. Ask me another?"

"DNA?"

"You must be psychic; deoxyribonucleic acid, a large helical molecule-the giver of life and heredity, first isolated by the *Society* in 1939, but formally discovered in 1944 from an extraction of cells."

"Jabot?"

"That's an unusual one. It's a piece of lace or cloth worn on the front of a neckband by men of the 18th century. Today, you'd have to be gay to wear one."

"All right, okay, I believe you. How can I not?"

"Would you like to hear more?" I asked.

"I want to know what you are, but no more questions," she said.

"Okay. No questions-just answers. Do you think you would believe me if I told you the truth about me now?"

"Absolutely, Christopher, I would."

"First, I want you to promise me that you'll never, ever tell another soul about what I'm about to tell you."

"I give you my solemn word," she said with her right hand in the air.

"I only wish you could promise me that you'd still love me afterward," I whispered to her softly. She gazed at me with puzzled eyes.

"All right. Here goes. I'm not a normal human being; never was and never will be."

"You look normal to me, Christopher, as a matter of fact, you're the most normal; better than normal, man I've ever seen."

"Darlene-I'm a clone."

"A what?" she asked with a quizzical smile.

"I'm a clone," I answered, showing her a very serious face.

"Christopher, that can't be true; there's no such thing. It's never been done with humans. Besides, the government wouldn't allow it."

"Darlene, it was done in secrecy; please, I'm telling the truth, and you said you'd believe me?"

"Okay, Christopher, let's say I do, but I don't understand how?"

"Then you do believe me?"

"All right. For the sake of your argument, yes, I believe you. Now tell me your secrets."

"Okay, Darlene, I'll start from the beginning. Here goes. I never knew my father, and I never had a mother. I was grown . . ."

"Hold it right there. Even a clone has to have a donor egg; therefore, you must've had . . ."

"You're not listening. The technique they used on me didn't require an egg. Believe me; the technology exists. My father had been dead for many, many years. He was a gunslinger from the old west-an outlaw. His body was discovered under some deep sand dunes; preserved as if he'd died only a couple years. And from his remains, I became the man that you know today, but with some exceptions.

I was raised on a military base; taught and trained to be the ultimate soldier. I didn't meet their exceptional standards and was given to the streets of this city when I was twenty-two. They gave me a fake identification and a very small wage to live on. After some years and pointless jobs, I found myself in that apartment. I suffered for about the last five years with terrible nightmares and headaches. I lived on alcohol, cigarettes, my wits and a gun."

"A gun!" Darlene exclaimed with concern, "Why did you need a gun?"

"For protection, and to help subsidize my income."

"I don't like where this is going; you sound like my ex-husband. What did you do with the gun?"

"I robbed people, Darlene, and I'm ashamed to say that some were innocence men. But most of them were just street trash; thugs, pimps, drug dealers, and other bad people who roamed the streets at night-preying on innocence people and others who were misplaced by misfortunate times."

"And by doing this, you thought you were better then them?" she said, showing a bit of anger.

"But Darlene, you saw the way I lived. It wasn't easy to survive in that neighborhood. I was even robbed myself!"

"But Christopher, you must've known you were doing something wrong. Have you ever heard the word; J-O-B!"

"I couldn't hold down jobs because of my mental condition. Something went wrong when they cloned me, something in my brain. It gave me horrible nightmares and headaches . . . God, if you only understood," I proclaimed holding my head.

"And who is this, *they*, you keep talking about?" she asked, irritated.

"The Society—a hidden secret part of our government with very intelligent people inside. They were the people who cloned me. They were the people who cured me of my nightmares and headaches. They were the people who cloned me again while I was alive. They are the people who are responsible for my strength that I just demonstrated for you. They are the people who gave me new hope, a new life, new intelligence, a new car, a new home and money."

"Okay, Christopher, I'm sorry-I didn't understand," she said sensitively.

I paced the floor for a couple of minutes thinking. We exchanged glances trying to decipher each other's thoughts and expressions. Our looks and feelings probably summed up to null. She looked sad, and I would feel strange. She then switched to compassion, and then I felt sorrow. Suddenly she started to cry just as I began laughing, and then her tears changed to a smile. I felt alone again. She bit down on her lip while staring at me with some kind of wonder, and I rubbed my face agonizing to resolve everything. We were both lost momentarily in a world without words. Finally, I started talking.

"Darlene, they've *re-engineered* me by modifying my genes. I've been mutated—transformed into a new breed of being. They called it: *genetication*. They told me I'd live another two-hundred years or more."

"Oh no! Don't tell me that . . . oh God please, tell me you're joking-please, and don't look at me like that . . . what about our future, what about me, what about our children and what about us?" she said crying again.

"Our children . . . Darlene?"

"Yes! Our children! There . . . I said it, I'm sorry," she said whimpering while brushing her tears away.

"Our children . . . yes, I like that. Our children. Then you still love me? Do you still love me, Darlene?" I asked on bended knee.

"Of course! I still love you, goddamn it!" she screamed into my face.

I grabbed her as she grabbed me, and we rolled on the floor hugging and crying, and then laughing softly from deep down inside. The moments compressed until we were kissing and taking turns wiping away each other's tears. We finally stood up and patted ourselves. We straightened our clothes, and then began holding hands while facing each other.

"Darlene," I said, bending back down on my knee, "Will you marry me?"

"Yes, Christopher, I'll marry you," she said with a warm smile and green eyes that shown nothing but love and sincerity.

Slowly standing, I hugged her gently, and then kissed her-it felt new all over again. I sensed a deep joy from within her, quietly, and yet slowing exploding, feeling her heart beat get stronger as the seconds went on. Our little courtship ended with a kiss and unison of the words: I love you.

I never told her I killed her husband.

"Christopher, it's getting late; would you take me home now?"

"Oh shit; yes, it's late and you have to get up early; I'm sorry. We didn't even eat."

"That's okay. I'll just make a snack before I go to bed; I'll live," she said with a twisted lip.

"Oh, one more thing before we go; I want to give you my manuscript." I went to my office and returned with a stack of papers. "Here it is," I said proudly.

"Hum . . . *The Calabi-Yau Universe.* Give me some time to read it and I'll give you an honest response."

Everything became more cheerful when we were driving down the road. On the way to Darlene's home, we talked and laughed like high school seniors planning for our senior prom-except our plans were a bit more serious. We contemplated when and where we would marry, what we would name our children. We decided to marry a month away, and for our children; if our first were a boy, we would name him Scotty, and if a girl, she would be named, Alyssa.

After we arrived at Darlene's home, I walked her to the door to kiss her goodnight.

"Darlene?" I asked again, "Will you keep my life a secret and tell nobody else?"

"I promise, Christopher."

"Good; and by the way, here's my phone number."

When I arrived back home, I went to my office where I read and worked on my computer. I surprised myself how fast I was learning. However, subjects that appeared to me to have a beginning and an end were the easiest to study: national history, surface geography, and normal human anatomy, including a few others. But I'm sad to admit, although most of these books eventually came to an end, at least to the present, there was too much room for discussion, argument and debate. Kind of like economics, philosophy, religion or politics. There seemed little room for absolute truth. Everyone had their own opinions, and still, everyone was right-no wrong answers or questions. Too much time wasted for me.

What interested me the most where the pure sciences; especially: mathematics, chemistry, physics . . . and all the others that leave you searching-waiting for the next break-through to soon be discovered. They were unlimited in our universe, but their answers concrete and solid. $2x + 3 = 7$, when $x = 2$, or $2O^{-2} + 4H^{+1} => 2H_2O$. And in the early twentieth century, $E = mc^2$. These were the truths that I desired.

I took it all in like a sponge. The fundamentals were just that-easy. It was the more advanced that I found challenging and stimulating. And there seemed to be no end to it. Everything became more complicated and abstract-even to the point of appearing to have no physical applications. But the pure sciences were like alcohol used to be for me-I couldn't get enough. Just when I thought I was reaching an end, a new book would begin with new theories, ideas and technologies. It became very clear to me, as many others have discovered: the more you learn, the more you realize what you don't know. And so on it went, as I did.

I had been awake for the last few days, studying, writing and reading. I kept in touch with Darlene, but not letting her know of my newfound ability-to stay awake for long periods of time. I also talked to Billy. He seemed to be improving. We talked and laughed on the phone as he told me how his store was becoming more of a reality. He had talked to construction, equipment, appliance and supply owners, explaining to them what he would need. Billy was still working from his hotel room, and I gave him another stay for a week or two. The bill was on me.

"Well Billy, sounds like things are coming together. I hope you're keeping receipts."

"Oh, yeah, Sandman, and your attorney, Dan Handson, has been helping me. He's even hired another attorney and consultant, *Alan Wasnick*, who specializes in large grocery store chains. Everything's ready to go. You should see it; the non-perishables will begin arriving tomorrow."

"Well I'll be damned, Billy, I'm very proud of you. It sounds like you've been very busy. I'm looking forward to seeing it."

"I've even taken the opportunity with a few local television stations to promote its grand opening, and I've also placed a full page advertisement in the newspapers asking for new employees."

"When do you start interviewing for positions?"

"Tomorrow. I would have called, but I didn't know how to reach you. But you did say you'd be with me when I started to hire; are you still planning on helping me?"

"No problem, Billy. I'm a man of my word. What time tomorrow?"

"Eight o'clock sharp! Alan will also be there to help choose the right people who understand the management of a large food business. Of course, I'll have the final say, but I want you there too, Sandman, I trust your judgment."

"How about I pick you up at seven-thirty?"

"I'll be waiting," Billy replied.

After a very restful and extended sleep, my house woke me up. "Christopher, time to wake up. Christopher, time to wake up. Christopher, time to wake up," that soft, female voice repeated.

"All right, already! I'm awake-what time is it?"

"It's five o'clock in the morning; your breakfast is warm and waiting. Shall I start a shower for you?"

"In thirty minutes, I need some coffee before I do my bathroom routine. How's the weather outside?"

"Partly cloudy, 52 degrees Fahrenheit, warming up to 72 degrees by noon with mostly sunshine." the female computer answered.

"House?" I commanded.

"Yes, Christopher."

"Select some clothes for me. Make them slightly informal, tan shirt, light brown coat, with the pants a little darker-no tie, and my tan leather boots please."

"Selecting," it said as I left my bedroom.

After a quick twenty-minute coffee and breakfast, I was in my bathroom shaving and brushing my teeth. Just when the toilet was flushing, the shower came on-a perfect 106 degrees. After eight minutes, I commanded the shower off and the dryer on. When I reached my bedroom, I saw that my clothes were extended from my opened mirror closet doors, including my boots.

At seven-thirty I was knocking on Billy's hotel room door. He was dressed better then me. I looked him up and down. "Don't forget, Billy, I get 10 percent of the net profits."

"I know. Dan-my-man has already taken care of that," he said, referring to my attorney.

"All right, let's go-I'll drive," I said politely.

"No way. I'll drive. That toy sport car of yours won't accommodate my wheel chair, but my van will. Don't think for a moment you're going to carry me out of here," he said jokingly and serious at the same time.

"You like your new van?" I asked curiously.

"Love it."

When we arrived at Billy's store I was spellbound. "God, Billy, this is terrific! You amaze me. It's perfect. Newly paved parking lot with plenty of lighting-and I'm especially impressed with those towering signs by the roads. I like it! Oh! And look at your logo on the store. I really like the way you've placed it on the archway with the bright and glowing, deep blue lights that sparkles: *BILLANCE*. That'll turn anybody's head around.

"I'm glad you like it, Sandman. You really help inspire my confidence," he said with the positive projection.

"This really is nice, Billy . . . let's go inside."

As we strolled through the big archway entrance, I observed a small line of potential employees waiting-maybe 25 or 30. If they were to be the extent, we were in big trouble. Once inside, I saw that everything was spotlessly clean and sanitary. New floors. The aisles were wide and spacious, and the shelving was built of polished chrome, each having its own hidden back-lighting. Different colored tubular lights hanging from the ceiling helped the shoppers know what area of the store he or she was in, or wanted to be.

And there were still the conventional signs that hung above each aisle spelling out the general items. Everything was computerized and itemized in alphabetical order, even within each aisle. But what really impressed me were the ceiling florescent lights: broad-spectrum lighting-the same light as sunlight.

"My God, Billy. This is incredible. How much did all of this cost yooou . . . me?"

"About six million. But I can assure you, Sandman, all of the freezers; floor and wall coolers are the latest designs and very energy efficient. I'm sure the store will be successful; I can feel it."

"Hell. I know it will. It makes all the others look like your old store. How expensive are you items?

"Cheaper by about twenty percent of the other super stores."

"How'd you do that?"

"Well, part of that six million, I used to buy another building-a storage tank if you will."

"Storage tank?"

"I buy all my non-perishables in much larger and cheaper bulks where I store them in my holding tank. And the shipping still cost me the same. I made an arrangement," he said with a wink.

"Billy . . . this better be on the up and up," I said with a look of suspicion.

"Oh! It is, Sandman, honest it is," he replied, and then pointed, "Look, there's Alan. What time is it?"

"Almost eight."

"Good. Now we can get started. Come on, I'll introduce you to Alan." I walked; Billy drove over to a table that Alan was setting up. While Billy presented me to Alan, I helped him with the table and chairs.

The more we got acquainted, the more I liked him. Through his eyes and body movements, I sensed honesty, integrity, openness and compassion. I might also add, that he communicated quite intelligently and freely. He was obviously a professional businessman, and a good one. I could see why Billy liked him.

As we were about to be seated, I noticed two security men standing at the front entrance. I then turned my attention to one of the large windows and almost shit my pants when I saw the line of people. They disappeared from somewhere around the store. Hundreds and hundreds of people! Billy showed me the advertisement he put in the papers, including what I thought was clever: the application. And all of those people had one, waving it around like the national flag. This indeed, was going to be a very long, long day.

"Okay, Steve," Alan said to one of the security men, "let the first three in."

"One last thing, Billy; don't call me Sandman; call me Christopher, or Mr. Lance. Understood?"

"Understood, Mr. Christopher Lance," he repeated.

And so we began our day; talking to strangers, asking questions, listening to many answers, making notes and changing lives. People of all walks of life were considered. Those who you knew that were living on the streets almost begging for work, and then others whose mother dropped them off for a job to do something for the summer. And the people that were just average, and many were, just looking for employment to make ends meet. In a way, it was heartening, like a barometric test, the pressure was constantly changing; some times high, other times low. Honest folks, lairs, truth tellers, story tellers, greedy, needy, cheaters, wife beaters, teenagers, retirees . . . the whole gambit of a small civilization sat down and talked to us. It felt like a marathon with short breaks.

At the end of the day, twelve hours, we had a stack of applications and resumes. Suddenly it struck me how serious this had become. On the table sat a pile of dreams, needs, wants, desires and goals-and lots of private information. Not so long ago, I wouldn't have given a shit, but now, I cared. And I wasn't sure if I liked that; there were too many people to think about. I was getting too involved. I wanted to return back to my new made life, with Darlene and my studies. I decided that from this point on, Mr. Billy Richers would be on his own. All he ever needed was just a lucky break, and I'm proud to say it was me who gave it to him.

In the late afternoon of the following day, I went to see Darlene. She still had her work clothes on when I arrived, but after about five minutes in her bedroom, she came out wearing a satin gown.

"Am I intruding?" I asked.

"Of course not, Christopher; would you like a cup of coffee? It's fresh."

"Humm . . . smells good-sure," I sat down at the dinning table. "Did you have a busy day?"

"It was okay; how was your day?"

"Good. I did some research on the network."

"What are you researching?" she said while facing me with two cups of coffee.

"Gravity. But I'm having some trouble understanding the math in Einstein's general theory.

"You mean the part about $E = mc^2$?"

"No-that was his special theory; that part was easy."

"Do you really understand that stuff?"

"Well, I'm beginning to, but I'm still in the process of understanding tensor analysis-actually, it's quite amazing. But what fascinates me the most is how mathematics is so intertwined with understanding the natural world. Math is everywhere-in every subject and every profession. It's the foundation of our existence."

"You know, Christopher, I like the way you're changing, and I'm truly in love with you, but it almost concerns me the way you keep on learning. When does it stop? Everyone has their limits."

"I'm not sure, Darlene; and I'm not sure if I want it to stop, but I have this insatiable desire to learn. For instance, did you know that everything is measured in just three simple units?"

"No, as a matter of fact, I didn't-should I care?" Darlene answered with disinterest, but replied, "Okay, I give up; what are they?"

"Mass, length and time. Everything can be measured with these. Speed, for example-length divided by time: (L / T). Acceleration is (L / T^2). Even energy is measured in these simple units: $(M\ L^3 / T^2)^{1/2}$."

"Why don't you go to college? Get yourself a degree."

"Yeah, what a joke that would be," I said smiling.

"I'm sorry if you found my question humorous," she said indignantly.

"I'm sorry Darlene, I'm not making fun of you; it's just that I feel college would hold me back. I don't care for their condescending attitude and regimental behavior. And they rarely understand anything beyond their specialty.

I'm afraid that most college instructors know very little outside their field of study. They specialize basically in their only knowledge and comprehension of usually only one field of study.

"You paint an ugly picture of these well intended scholars."

"Sometimes the truth hurts, but I will admit, there are always exceptions."

"But Christopher, that's the purpose of taking many different classes-to give you a well rounded education. And you can't expect one person to know everything."

"But I learn much more, and faster, by reading specific subjects by many different authors. And besides, I don't have the patience of wasting time on exams and quizzes."

"Well, please correct me if I'm wrong, but those exams and quizzes are to help reinforce your understanding, and to give the instructor some feedback on how well you're doing."

"Darlene, I'm not trying to be rude, but within the past month I've learned more than taking four years of college. I can read faster, and comprehend what I read much more quickly then making oral lectures on the same subjects that are already in books, or the computers. I don't need a degree to get money-what does it matter if I'm self taught. Should I need a person or a group of people to hand me a piece of paper telling me what I should learn, or what not to learn, or telling me whether I'm smart or not?"

"You see, Christopher? When you talk like that it bothers me, besides, how can you be so sure-sometimes it helps to work in groups."

"Here we go again-you'll just have to trust me. You know, there's also a little known thing called: The Scientific Method, more commonly called: research and development through experimentation."

"You perform experiments?"

"As a matter of fact, I do. I have an elaborate and multifaceted laboratory across from my office."

"Christopher, can we change the subject?"

"Am I bothering you?"

"A little. Would you like another cup of coffee? I'm having another one."

"Sure . . . but you stay put-I'll get it." I got up quickly to help change the subject. "I know you like yours black," I said while making the two cups of coffee. I placed the cups slowly on the table while watching Darlene stare into space. "Whatcha' thinking?"

"What do you think?" she said smiling.

"I'm not sure? Our marriage and children?"

"Kind of, in a roundabout sort of way, Christopher. But I have to keep asking myself, do you really love me?"

"Of course I . . ."

"I mean, do you love me enough to stay with me for the rest of my life?"

"Oh. I see. You're worried about getting older-older than me."

"Funny, I think I'd believe you if you told me you could even read minds. Have you thought about it?"

"Yes. I've thought about it?"

"And?"

"And . . . I'll have to adjust."

"Be serious, Christopher."

"I am," I said tying not to think about the future.

"You'd want to fuck an old lady?" Darlene said with a bit of anger.

"Darlene, you know it won't be that way. Life will always have its own rules no matter how long we live."

"But what about our children? What will they be like?"

"Time, sweetheart; we'll just have to wait and see.

"And you still want to marry me?"

"With all my heart, Darlene," I said in the most reassuring tone. She got up, said nothing, but sat on my lap and kissed me while I did the same to her.

Softly and slowly, she got up pulling my hand gently with hers', "come on . . . let's have some wine and soft music." I followed her to the sofa. Darlene was pouring two glasses when her phone began to ring.

"Hello?" she said politely. "Hello?" she said again. I looked at her thinking it might be an adulterated call, and then she turned to me and said, "I think it's for you."

"Hello, this is Christopher speaking . . ."

A serious voice, which I knew, was on the other end. "Hello, Mr. Sandman, we have some important business to discuss. See me within the next two hours." *Click.* The phone went dead.

"Who was that? You didn't even say one word. Are you in some kind of trouble?" she said looking concerned.

"No. That was just my boss; I have a meeting to attend."

"Why so late? Is it that important that it can't wait till tomorrow?"

"I'm afraid not, Darlene. I'll have to be leaving soon."

"It was from that Society you told me about-wasn't it? What are they going to do to you?—More experiments? Please be careful."

"Oh, it's just business-nothing to worry about," I told Darlene, and hoping what I said was true. "Come over here and give me a hug." I sensuously licked her ear and then whispered to her, "I love you."

On my way to the Society, all I could seem to think about was how Scott knew where I was and what he wanted to talk about. I concluded he had the doctors' surgically implant a homing device-a GPS somewhere in my body. And after more time thinking, I was quite sure what he wanted to talk about. Now the only questions that remain, was how much trouble I was in and what he would do about it. Would he have me killed? I doubted it. Would he give me a tongue-lashing? Probably. Should I care? Most likely, after all, Mr. Scott Sellers was the point man at the Society.

The wind felt good and warm and just the right temperature-a lovely summer night and a dreamer's-wish come true. And to think I was driving through it to get a spanking. Wasted time.

I decided to quit thinking about it, and instead, take in the evening under the stars. The night was nice in the mountains; I could see the stars from my left, and the city lights down below from my right. As I drove higher up the mountains, and with time growing longer, I was starting to see the building. It looked larger then the last time I saw it, and it appeared to be more spread out. But I can't remember it looking so ominous. I took a deep breath as I approached the opening gates.

I pulled up to the main courtyard in front of the doors. When I waved my card through the slot, the doors opened. I walked in the lobby and was relatively surprised to see Scott and Bradly standing at a distance in the foreground.

"Welcome, Mr. Sandman . . . come with us; we have some issues to talk about," Scott said impersonally. "We'll go to my office." When we arrived in Scott's office, he poured me a drink. "I take it you still drink, or have you quit, Christopher?" Scott said sounding quizzically interested.

CHAPTER SEVENTEEN

"I just drink for the taste now-I'd have to drink a gallon or more of this shit if I wanted to get drunk." I simply retorted.

"Is that so?" Bradly answered.

"Sit down, Christopher-would you care for a smoke-or have you quit those too?" Scott said.

"I just smoke . . ."

"Yeah, yeah; for the taste I suppose," Scott concluded.

"Why am I here, Scott?"

"I have some important information to tell you, and I wanted to tell you now-I didn't want it to wait any longer?"

"Okay, I'm focused; speak to me."

"Mr. Lance?" Bradly surprised me when he spoke. I shifted my attention as he continued. "We have you *chipped*."

"Tell me something more that I haven't already figured out. You must think I'm stupid. You have a GPS in me . . . and let me guess; you need to take it out now. Am I right?" I said confidently.

"A very good conjecture, Christopher. How did you know we wanted to take it out?" Scott replied.

"Because of the 8 to 10 *sieverts* of radiation exposure leaking from this damn thing planted in the back of my neck-I couldn't feel it at first, but now it's becoming quite uncomfortable. Using a Geiger counter, my calculations confirmed it. You know and I know, I'll die if you don't remove it soon."

"You're good, Sandman, but I shouldn't be surprised. So tell me, when are you and Darlene getting married?"

That question made me stand. "How in the hell could you know that-oh, how stupid of me-it's this goddamn bug you planted in me. You could also hear my voice?"

"Quantum physics, electron spins, and Tellurium chips with nanotechnology and GPS. We knew your location, the words you spoke and heard. We even knew your heart rate and blood pressure among other things. Hell . . . we even knew when you took a shit. We'll take it out before you leave, Christopher, but I also have more news to tell you," Scott said while crossing his legs.

"Is it good or bad, Scott, what are you hiding from me now?"

"Well, initially it's bad, but eventually good," Bradly said.

I turned in his direction waiting for more. He turned his head toward Scott and said nothing more. Sitting back down slowly and turning toward Scott, I waited for words.

"Christopher, when you were re-engineered, we did something to you-for precautions of sorts."

"Hold it, Scott, don't say another word-let me see if I can figure this out using my special powers you've given me. I'm just going to sit here quietly for a moment and let my body sing to me."

"Come on, Sandman, quit screwing around . . ."

"I've figured it out. I'm the first human with only 44 chromosomes. You removed my sex chromosomes. Am I correct?"

"You're thinking too deep-we simply gave you a vasectomy, reversible of course. It'll only take a couple hours to reverse your vasectomy and remove the bug. You'll be able to leave tonight."

"Is that it?"

"No. I have a few other things I would like to discuss with you before you go under the knife." Scott said rather seriously.

"May I have another cigarette first?—Thank you Bradly. Okay Scott, I'm listening."

"After we remove the implant and unclamp your vas, I need a sperm collection."

"Why?"

"I think the reason is obvious," Scott replied.

"You want my DNA? Why don't you just draw some blood?"

"It's not that simple, Christopher. I want your sperm cells. A sample of blood would require a substantial amount of time and energy to extract your x and y-chromosomes. Why bother with blood cells when your sex chromosomes can be had without complicated measures?"

"You've got something that I feel you're hiding from me."

"Not really. Do you love your girlfriend?"

"You know I do."

"Do you want to watch her grow old while you stay young?"

"Of course not, but we've already discussed that."

"And what about your offspring, what will they be like?"

"They'll be like other children. Just like . . ."

"Wrong, Sandman. The only thing Darlene will give birth to will be a glob of cytosol. Your sex chromosomes will harm hers. You have to remember, Sandman, you're a new species. Your sex chromosomes are far too different from hers."

"So what's your point-why are you telling me this?"

We want to mutate Darlene. We want your mate to be compatible in the same manner as you."

"Come on, Scott; do I look that stupid to you? You know as well as I, that it'll take years to sequence her genes. Large genomes, like human beings take . . ."

"Christopher . . . shut up and listen. When it comes to genetics, you're light-years behind the Society. We have advanced tremendously. The books and other materials you've read about genetics are out of date. We have an Automated Sequencing System that can process massive amounts of DNA. Aside from our front-end procedures, we've developed the latest in technologies for our back-end procedures as well: our thermal cycler, sequencer and plasmid purification devices are unknown to anyone outside this complex. We have cubic, nano-computer processor units and machines that can rapidly automate editing, assembly, sample tracking, and I may also add, with perfect quality control; and we're still making advances. Our IDNASS (Integrated DNA Sequencing System) can handle, thus far, 1 GB per hour. That's 1,000,000,000 base pairs every hour-have you read that in any articles or books?" Scott finished, and then drew a big breath.

"Absolutely not!" I replied in astonishment.

"She will become like you. She will live as long as you, and your children will be nothing short of a miracle. Do you follow what I'm saying?"

"Yes, I do. We will be a quantum leap in punctuated evolution-we will be new human species. What would you call us? Oh . . . I remember now; Homo genesis sapiens-like a modern version of Adam and Eve."

"That's an eloquent way of expressing it, Sandman. So what do you think? Do you think she'll be willing to change for you?"

"I'm not sure. I'll talk to her tomorrow and get back with you. Buy the way; you don't have any more of those spying bugs implanted in me-do you?"

"No. Are we square on what we have to do?"

"We're square, Scott."

"Good. Now off you go to the infirmary, Karen is waiting for you. I'll expect a call from you tomorrow, after you talk to Darlene?"

"It'll be later in the evening. Give me some time to confer with her."

It was nice to see Karen again. She's a very good lady. She almost had the appeal of what I surmised as a mother.

"Well, well, Christopher, here we meet again. How have you been feeling?"

"Never better. What you folks did was incredible. The more I read, the more I learn, and I have a lot of reading to do yet. I've amazed myself of my ability to read very quickly, and my capacity to comprehend and advance to higher-level topics. My physical aptitudes have also improved. I'm even beginning to hear my body sing to me. I know that sounds strange to you, but not to me. So, Dr. Karen Simson, how have you been?"

"Very busy, Christopher. How is your pain threshold?"

"I don't know-'bout the same I guess if something hurts me. A couple weeks ago, I accidentally dropped a filing cabinet I was carrying and it fell right on me toes. That hurt like hell."

"How many compartments were in the cabinet?"

"Four."

"Were they empty?"

"No. I was trying to save time."

"Okay then, I'm going to numb your neck so I can remove that damn bug they put in you. I knew those tiny nuclear batteries where too strong, and with tendency for leaks."

"Will you do me a favor, Dr. Simson?"

"What's that?"

"Don't numb my neck, I want to try something."

"What?" Dr. Simson asked.

"I want my body to sing to me so I can focus."

"Focus on what?" Karen asked with a small smile.

"The back of my neck. I want to see if I can block out the pain. Just give me a moment before you start cutting."

"It's been thirty seconds, can I start now?" Karen said.

"Cut away."

"Do you feel any pain?"

"No, but I can feel the microscopic ripping of flesh, and the warmth of the blood that surrounds the split. Well I'll be damned; it works. Another interesting discovery I've learned about myself," I said amusingly.

"I'll be finished in just a moment. I have to glue it and patch it, and then we can work on reversing your vasectomy," Karen said with a blank face.

"If you don't mind, Dr. Simson, this time I think I'll let you numb me."

"Okay. I want you to lie down on this bed on your stomach with your knees near the edge. Now lift your legs and feet up, and then spread them slightly. Good. Now, with your hand I want you to move the shaft of your penis pointing due north, and flat under your stomach. Now push your scrotum as far back behind you as you can. Can you nudge yourself back just a little bit more? Very good, now I'll numb you."

I could feel the latex covered fingers probing and sticking me. After a few minutes, even my dick went numb.

"This won't take long." Karen said.

I felt some pressure and movement, but no pain. After about forty-five minutes, Dr. Karen Simson had glued the incisions closed. She put some kind of patch on my scrotum, and then told me I could put my clothes back on.

With some hesitation, I told Karen, "I'm not sure if I can give you a sperm collection now."

"No need to," she said plainly, "I took some before I reconnected your tubes."

"Will it hurt after the numbness wears off?"

"There may be some discomfort for a day or two; but nothing I'm sure should really bother you. Oh, and after you dress, Scott wants to talk to you. Goodbye Christopher." After I made my way to Scott's headquarters, he was there to meet me.

"Please, come in Mr. Sandman, I have some interesting news and a proposition to offer you."

"Talk to me O mighty master of mystery?"

Scott looked intense but self-assured. "For the last time, and I want to hear it again: do we want to geneticate Darlene? Because of our advances, it should take less then a month, but when we're done, she'll take on the same attributes as you."

"She'll live longer?"

"You know for yourself, Christopher, that's only part of it-do you think she'll go for it?"

"I hope the hell she does . . . this is a wish come true!—But wait; why?"

"To begin the new race," Scott simply replied.

"To do what!" I proclaimed.

"Don't ask me again. I'll explain later. Go and talk to Darlene." As I turned to leave, Scott said one last thing to me.

"Remember, Christopher, this is classified. Tell no others but Darlene; are we clear on that?"

"We're clear," I said as I walked away and unsure what to think.

"Are you serious!" Darlene yelped.

"I couldn't believe it myself-do you know what this means, Darlene?"

"Yes . . . no; oh shit, I don't know-I'll have to think about it."

"What's there to think about?"

"I don't know, Christopher, but I'm scared."

"Don't be afraid, Darlene, I wouldn't put you into a position that I thought would harm you. You have to remember; I was their first. I was created from a soup of my father's genes; and then they re-engineered me. I am a new, wonderful being. I am a new race and I want you to follow me. Please baby, trust me."

"But what are you?"

"They call me Homo genesis sapiens . . . I am becoming genius, I am supreme among the others; I am truly a new breed.

I can be fearful and angry, but I can be very gentle too. I can think faster now, and I can see more keenly than a bird. I'm also beginning to see in different spectrums. I can attack with lightening speed-I can kill for my safety and for those around me. I am an ultimate new quest. Please, Darlene, don't forsake me; join me."

"And if I should say no?" she asked.

"Then what you decide goes."

"Would you leave me?"

"Of course not; just the opposite, and it breaks my heart."

"I don't understand, Christopher. What are you trying to say?"

"I'm afraid you'll want to leave me when I tell you some new news."

"What news?"

"When we first met, I was sterile but I didn't know it. I just had a reverse vasectomy, but unfortunately, we can't have any babies."

"Well . . . that doesn't make any sense? You can ejaculate sperm-can't you?" Darlene said looking totally confused.

"Yes, I can, but our gamete cells are incompatible."

"Speak English, Christopher."

"Our germ cells are not compatible for mitosis. Your egg cell and my sperm cell could never form into a fetus. This has nothing to do with you, but everything to do with me. I'm very sorry, Darlene, but I can't give you any babies-unless . . ."

"Unless what, Christopher?" she said compassionately.

"Unless you undergo the same type of re-cloning, or genetication procedure as I did, then and only then, could we have babies. Our children would be very special too, and in a good way. They would live, and you would live, a comparable longer life like me."

"So what you're telling me, Christopher, is that if I undergo this cloning procedure, we and our children will become this new breed you spoke of? What'd you call it?"

"Homo genesis sapiens," I said as a matter of fact.

"Is this good, or bad, Christopher? Please . . . oh please, be honest with me," Darlene pleaded.

"I've never felt better in my life, Darlene. There are no bad side effects. I love my new life just as much as I continue to grow more in love with you. I feel absolutely great!"

"How long do I have to think about it?"

"I'll have to know by the end of the day. There's no time-window on this thing-it's either yes or no. I'll call you later when I'm home."

"Yes," Darlene softly spoke.

"Yes, what?" I asked.

"I'll do it."

"Are you sure, sweetheart? If you would like a little more time to think about . . ."

"I don't need any more time. I love you, Christopher, and I want us to grow old together. I want to marry you and have your babies.

"You've made me so happy, Darlene, I could cry," I said with watery eyes. And then I hugged her deeply and she in turn replied.

As we stepped back away, facing each other and holding hands, she softly asked me, "What do we do now?"

"I'm going to call Scott for the details-he's waiting for my reply." I picked up Darlene's phone and called a secret number.

"Hello, Christopher," Bradly answered, "Would you like to speak with Scott?"

"Please," I said.

"Hello, Christopher," Scott said, "What's the word?"

"Darlene said she'd do it-when do we start?"

"Bring her to the Society . . . now," he said, and then I heard a click.

"What did he say, Christopher? That was a very short call."

"He said we must leave now."

"You mean, right this moment?" Darlene said with a trace of panic. "What about my job, my rent, my clothes and other things; what about my other obligations and expenses?"

"Forget them; they'll be taken care of. After the procedure, I want you to stay and live with me . . . marry me. Everything will be okay. Sweetheart, please have faith in me.

"Well, at least let me pack some things."

"Don't bother; the Society will provide anything you need or desire."

"Okay, alright. Jesus, I can't believe this is really happening to me," Darlene said apprehensively.

She locked the door to her apartment, and then followed me to my vintage, modified bulletproof Jaguar XJR-9LM. "Don't drive like a maniac," she said sternly, "I'm in no mood for dying today."

"We'll take it nice and slow," I said insuring her.

"How far away is this place?"

"About a two hour drive at this speed."

"Alright, drive at your speed, the sooner we get there, the better I'll feel-I hope."

After we had left the city, I could have driven the roads much faster, but I decided to meet her halfway instead of my desire to go faster. It was becoming dark outside when suddenly my headlights lit up. Ninety miles per hour down the black empty roads didn't appear to frighten her. Finally we reached the outer main gates and I stop momentarily for them to open. Like gates to a foreign and forbidden land, they slowly and mysteriously swung open. Darlene's eyes were wide and white with wonder and trepidation.

As I made my way to the Society's complex, Darlene seemed a little concerned when she saw the shadows of soldiers and their weapons patrolling the road. When I parked near the main front doors, Sergeant Cooper was there to meet us.

"Nice to see you again, Mr. Sandman," he said with a quick salute.

Darlene looked at him and then at me. "Who's Mr. Sandman?"

"I'll explain later," I said to her, and then set my attention on Sergeant Cooper. "Show me to your leader?" I asked in jest. And without a smile, he turned around and slid his I.D. card in the appropriate slot. The front doors parted and we walked in.

"Please follow me," he said as we walked into the main floor lobby. "Have a seat here; Scott should be down shortly."

After a few minutes, Scott walked in sporting a white, short beard and a handsome haircut; wearing an expensive tan suit with polished clean boots. He approached and stood before us. When he looked at Darlene, I could easily see that his presence intimidated her. His persona gave away his intelligence, wealth and power.

"Good evening, Christopher. It's a pleasure to see you again. Ah . . . and this lovely young lady must be your fiancée, Darlene?" Scott politely spoke and extended a hand shake. "Please, let us go to more comfortable surroundings."

We followed Scott to that familiar elevator door, which Darlene had never seen before. "Take us to level four, Christopher." I pushed the elevator button and it lit bright green as it read my fingerprint and blood type, and then it took us to my temporary apartment where I had lived before. To my surprise, everything had changed.

There was new carpeting and new furniture, and different colored walls. The living room appeared to still have the huge screen TV because I saw that same remote resting on a stand. I was engrossed with the new surroundings, when suddenly a voice broke my glances.

"Are you still with us, Christopher?" Scott said good-humouredly.

"Oh, I'm sorry-I was just admiring the changes."

"Please . . . Darlene, have a seat. May I get you two a drink?"

"Would you have any sherry?" Darlene asked.

"Let me see," Scott said as he bent slightly to look under the bar. "Would you care for a glass of amontillado?—I've heard you like it."

"That would be nice," Darlene answered.

"What's your pleasure, Christopher?" Scott asked me as he was pouring the glass of sherry.

"I'll have a taste of scotch," thank you.

"On the rocks?" he inquired.

"Sure," I answered simply.

Darlene and I were sitting in a love seat. I had my hand on her lap and hers was on mine as we watched Scott set the drinks on a coffee table in front of us. He slowly walked back to the bar, poured himself a drink, and then sat across the room and began to speak.

"Welcome to the Society, Darlene."

"Thank you," was all she said.

"I trust that Christopher has spoken of us-nothing bad I hope?" he replied.

"Oh no . . . I mean, yes, err . . ."

"Say no more, Darlene, I understand, and I see nothing but good in your eyes."

And with that said, Scott picked up a phone and pushed a button. He said some words that I'm sure Darlene couldn't hear, but I could. We would be expecting company very soon.

"Would you care for another drink?" Scott asked while exchanging looks at the two of us.

After a silent moment, Darlene interposed, "Yes, please, I'll have another glass if I may?"

"I'm fine," I replied.

As Scott was pouring another drink for Darlene, I heard the elevator doors open with footsteps approaching. "Jolly well, everybody's here," Scott said smiling. "Please, everyone, feel free to make a drink-there's plenty of everything. Afterward, take a seat and I'll do the introductions."

In the room, there were seven people besides Darlene and myself. To my disbelief, I swear I even saw Billy without a wheelchair. There were also two other men in the room, which I hadn't seen before.

I watched everyone make their drinks as they talked softly to one another. Darlene and I looked at each other and I could tell she was just as confused as me. Eventually, everyone sat down and Scott began to speak.

"Tonight we are gathered here for a special occasion and celebration. As you may all be aware, Mr. Sandman is the one that made it all happen," Scott said as he held his glass out in my direction. "He was, and still is, our inspiration. He was the first, and the alpha of our new generation. He was the one who helped us prove it could be done. He set the example, and then we followed. And tonight we have another. Everybody; I'd like for you to meet his future wife, Ms. Darlene Osmon. Would you mind standing here beside me, Darlene . . . you too, Christopher?" Scott spoke with a grin from ear to ear.

As we stood to accompany Scott, Darlene whispered in my ear, "Who is this Mr. Sandman?"

"It's me," I replied quietly, "now, let's just walk over to Scott and I'll explain later." She followed my instructions and before we knew it, everyone was standing while Scott was starting the introductions.

"Christopher, some of these people you already know, but Darlene has never met any of them. Let me start by formally introducing myself, Darlene. I am Dr. Scott Sellers. But please, just call me Scott." Darlene looked at him as he slightly bent his head forward, showing her his respects.

"And to my left, standing over there, is Dr. Bradly Hues." Scott said with his glass extended again.

"It's a pleasure to meet you, Ms. Osmon," he said bending his head slightly to mimic Scott.

"And next to his left, I'd like you to meet Dr. Karen Simson," Scott announced as she followed accord.

"And next to Karen is Dr. Wayne Koloski."

"Dr. *Latex* Koloski," I mumbled to myself.

"I'm sorry, Christopher, did you say something?" Scott genuinely asked me.

"Ah . . . no, I was just clearing my throat . . . please, continue."

"Standing to the left of Dr. Koloski are doctors', Vance Ringrose and Otto Betenkulf," Scott said, and then added, "and finally we have, Mr. Billy Richers."

"Billy!" I bellowed, "Is that really you?"

Billy broke his fake straight face and burst out laughing, "Yes, Sandman it's me; it's really me!" he cried out as he walked toward me for a big hug.

"You're walking! And where did your beer belly go? You look so healthy-what's happened to you?" I replied in complete amazement.

"The same thing that happened to you, Christopher," Scott said proudly.

"You mean, geneticated?" I had to ask, but had suspected.

"Everybody in this room has been geneticated, with the exception of Darlene, of course," Scott replied straightforwardly.

"Everybody!" I said loudly, and feeling utterly flabbergasted. "How long?"

"Just a few weeks," Bradly responded.

"Okay, yes, I understand that, but . . . how long?" I had to reply.

"I'm just a bit confused, Christopher; can you be more specific?" Scott asked while Darlene just stood silent-wide-eyed and bewildered.

"How long is everybody going to live?" I asked explicitly.

CHAPTER EIGHTEEN

"Oh . . . well, I see. That depends on your present age," Scott answered. "Now, take me for example. I should theoretically live for about another 75 years. Bradly has approximately 100 years. Otto and Vance have about 125 years before they die. Karen and Billy may have about 150.

You, and I presume soon, Darlene, will enjoy about 200 more years. The mathematics is rather complicated. They're exponential perturbation formulas that require computer analysis. You'll understand their meaning as time goes by."

"How about strength, eye-sight, hearing and all the other things I'm experiencing?" I asked Scott seriously.

"Well, Christopher, these are also determined by similar exponential perturbation functions. And again, your present age at the time of being geneticated, will also determine those things. The older you are when you're geneticated, strength, eye-sight, intelligence and other such attributes will be less intensive than those who are younger. You'll understand these things after time." Scott repeated again.

I turned to Darlene and said, "You heard the man; do you still want to be geneticated, like me and the others?"

"Geezzz . . . it all seems so strange now. I feel like I'm surrounded by a bunch of aliens or something. Are you people still human?" Darlene turned to ask Scott.

"Well," Scott replied, "in a queer sort of way . . . yes and no, but mostly no, we're not."

"Is this a good thing, or just some sort of weird experiment?" Darlene turned around and asked me.

"Well, sweetheart, to be completely honest with you; I'm not absolutely sure . . . but it feels like a good thing to me. Look at Billy. Just a few months ago he was in a wheelchair, paralyzed from the waist down. And if you're geneticated, you'll live as long as me; and think about our children. Scott?—What about our children?"

Scott cupped his hand it front of his mouth to cleared his throat, and then he spoke, "I'm not a betting man, but if I were, I'd bet your children would be absolutely astonishing-geniuses by the age of ten, and far ahead of Homo sapiens. Like us, your children will be immune to diseases, and will never get ill. They should take on all the attributes that Sandman has, but most probably to a greater extent. And I'd have to do some calculations, but my guess is that they would live to a ripe old age of 400 years or more."

"They would be freaks!" Darlene shouted in an inelegant way.

"Darlene . . . please, relax and listen. Your children will be very special and very gifted. It will be your children, and many others like them who will journey outward and take us to the stars and other worlds. We are on the threshold of a new evolution with so many new discoveries. I feel very strongly in the direction I'm taking. I don't know how to explain it, but I know it's the other part of the human equation," Scott said very calmly and softly.

"Is that all there is to it?—Everything is nothing but equations and numbers?"

"Of course not, Darlene. Look at Christopher; look him in the eyes and tell me you don't see emotion. Tell me that this man doesn't love you deeply. Has he done anything to hurt you? Does he not seem human to you? Think about it-your children would still be God's children like any others," Scott replied with his arms extended downward and his palms facing upward.

"And what about God? Is this what He wanted? Isn't this stuff sinful?" Darlene asked and pleaded.

"I don't think so, Darlene. Our Creator gave us freewill and intelligence, at which He allowed us to discover His secrets and mysteries. Think about this, Darlene?—In all our surroundings, which extends outward throughout the universe, there is matter and energy-there is something. What if there was nothing? There are only two choices; something or nothing; and what do you see or perceive?"

"I'm not sure I understand what you're saying?"

"Okay . . . let me back up a little. I shouldn't have mouthed the word, "God," in this discussion. I should leave that up to the Holy Bibles and theologians and in the realm of philosophers; but I'll agree that it's hard to ignore. Whether this stuff is sinful or not, I don't know what to tell you except that it's real and its happening. I do believe if genetication were evil, it wouldn't have gone this far-I believe good always overcomes evil. Now, let me ask you a hypothetical question. Suppose you had a child who grew to an adult but was stricken with a defective heart that would soon be his or her demise without a transplant. And let's assume that doctors found a heart donor and were willing to replace the defective heart with a good one; would you be willing to do it?"

"Of course I would," Darlene answered.

"Okay . . . good," Scott said, and then continued, "but wouldn't this medical procedure be an intrusion on God's will?—After all, your child was destined to die without medical help. Is what we're doing any different? All of us in this room will lead better, healthier and longer lives. Where is the harm?—You tell me because I don't see it? Sandman . . . talk to her because I don't know what else to say."

"Why do they keep calling you, "Sandman," Christopher?"

"Bradly?—Would you please help me here and explain why I have this nickname?" I asked with exasperation in my voice.

"Sure," Bradly calmly said, and then sat down with her and explained the whole story while everybody refreshed their drinks, and then sat down to listen. After about twenty minutes, Bradly was concluding, "And when we realized the success we had with Christopher, we knew the benefits of geneticating, and proceed with ourselves. You may ask anybody in this room if they regret being geneticated. We'll start with you, Christopher-do you have any regrets being geneticated after you were cloned?"

"No," I answered, and then exchanged places with Bradly so I could hold Darlene's hand and look her straight in the eyes. "Sweetheart . . . I know all of this must seem very strange to you; probably like an eerie dream or something, but baby, it's happening, and it's real. I've never felt better in my entire life. I'm beginning to understand things that I would have never understood before. I can filter the spectrum and see in many different colors and I've already demonstrated my strength. Many wonderful and incredible good things are happening to me. My senses are becoming much more acute and stronger; for heaven's sake, Darlene, I'm beginning to "hear" my body sing to me, and it sings such sweet songs-try and explain that to someone-but it's all good sweetheart, you just have to put your faith in me and believe that I'm telling you the truth. I would never ask you to do anything that I thought might harm you. Being geneticated is the obvious next step to a healthier life. Please, Darlene, say you'll do it and I'll assure you that you'll never live to regret it . . . what do you say, sweetheart?"

"Okay . . . I'll do it."

"What?" I had to ask.

"I said; I'll do it. I love and trust you. I don't feel threaten by anyone here, and the more I hear about genetication, the more I'm beginning to like it-I want to live as long as you so we can have children and grow old together. I'll do it," Darlene ended her sentence with a soft and pleasant voice.

"Oh, boy, Darlene; you have no idea how happy you've made me. I love you and always will," I told her gently, and then turned to Scott and the others and said," Okay, people, where do we go from here?"

Dr. Karen Simson spoke up and said, "If you don't mind, Christopher, we would like to explain the procedure to Darlene and show her around a little so she can feel a bit more at ease. We'll be back in an hour or so, and in the meantime, why don't you and Billy get re-acquainted again. I know it's been a few months since the last time you two spoke together," Karen said with a smile, "and don't worry about Darlene; we'll take good care of her."

After everyone left, Billy and I stared at each like adolescents, who for the very first time, were without supervision. I smiled at him and he smiled back, and then I started to laugh and he did the same. Billy walked up to me and grabbed me by my shoulders, and then gave me a big bear hug while lifting my feet of the ground, swung me around and sat me back down-he surprised me with his strength.

"They did you; didn't they?" I asked.

"You're damn right! I've never felt this good, and I owe it all to you, Sandman."

"They even fixed your spine."

"A week after the procedure, I was out of bed and walking. I have to tell you, Sandman, it's like a gift from God. I truly feel blessed."

"Well, I'm very happy for you, Billy."

"Have you seen my store lately?"

"No, I haven't-how is it going?"

"Haven't you checked your monthly statements in your *Business Money Market Account*?" he said clasping he hands.

"No, I haven't."

"What? You don't read your mail? Well, I think you'll be surprised. You should have at least, $150,000 in there-my business is doing great!"

"That's good, Billy. I'll have to stop by BILLANCE and take another look sometime. Where are you living now?"

"In a nice two story colonial home not very far from my store-Scott bought it for me."

"I'm curious, Billy; how is it that you became geneticated?"

"Well, it's kind of a strange story, but this is basically what happened. As you know, I was staying at the Emerald Hotel with I heard a knock on my door. I was surprised because the desk clerk didn't call me to let me know I had company-I thought it was you again; maybe coming back for something you had forgotten. I didn't say anything, but simply opened the door and there they stood.

They walked in like soldiers wearing black suites and shinny shoes. "We'd like to ask you a few questions, if you don't mine of course?" the taller guy said in a deep voice. At first, I thought they were FBI men, but something didn't seem quite right. Then they asked me about people I knew and when your name came up, they hauled my ass away. I was carried to a long black limousine and taken to here, right where we're standing. Damn, it scared the hell out of me. But then Scott came in and introduced himself, and then explained what it was all about and what he was going to do. So in a corn nutshell, well, here I am. Do you want another drink?"

"Yeah . . . I still like the taste of good scotch, but I'd have to drink a lot to get me drunk."

"Funny you should say that," Billy said, "it doesn't seem to have that much effect on me too."

"You know why; don't you?" I said expecting he knew.

"Because I've been re-engineered, or geneticated. That's what healed my back and also gave me strength."

"You haven't seen the last of it," I responded, "give it a few more months, and you'll be surprised at what you'll be able to do."

We talked more about Billy's store and what I was doing for a living.

"Well, Billy, I've been doing a lot of research and learning while writing."

"Really?" Billy said and continued, "What are you writing?"

"Encyclopedias," I answered with a smile.

"Was that a plural I heard?"

"You heard plural," I retorted, "Right now I'm in the beginning stages of four different encyclopedias: one on Mathematics, the second on Physics, the third on Chemistry and the fourth on Cosmology."

"Why?" was Billy's only response.

"First, because I can and second, because I want too and third because I promised Scott I would. As a matter of fact, he's already prepaid me a good sum of money for my works."

"To each their own private glee," Billy simply said.

We were starting a new conversation when unknowingly, Dr. Karen Simson and Darlene walked in.

"We're back, Christopher. Darlene is going to stay here for awhile; if you wish to stay with her, you may, or if you want to go home, you may do that too-it's up to you," Karen politely said.

"How are you doing, Darlene?" I asked.

"Oh . . . she's fine. I just gave her some pills to settle her down and help her relax," Karen said with her arm gently wrapped around Darlene's shoulder and the other hand embracing her arm.

"How are you feeling, Darlene?" I asked again.

"I'm okay, Christopher . . . Karen explained everything to me and I'm satisfied that I'll be all right. I'm actually looking forward to the genetication process so we can get on with our lives."

"Would you like me to stay with you tonight, Darlene?"

"You don't have to and I'm not in a very talkative mood. I'm feeling a little tired and the procedure starts early tomorrow." Darlene calmly said.

"Will she be okay, Karen?—Or should I stay?"

"You may leave, Christopher; you know she'll be safe." Karen said reassuring me.

"Will she have to take that sanitizing shower in the morning that will remove all of her hair and the plugs to . . ."

"No, no, Christopher. We've made monumental strides in the genetication process. We have computers now that make computations in a matter of minutes instead of days, when we geneticated you. However, she'll still take a sanitizing shower in the morning but she won't be losing any hair. And there is no more need for cavity plugs anymore. Like you, we'll start with her brain, but the new needles we use now are so small that they'll drill between the follicles of her hair. And she won't have all the little, red prick marks that you had. The procedure now takes only a couple of days instead of a week. She'll be off the operating table and in bed in two days.

As a matter of fact, we've already drawn her blood, and her complete mapping and sequencing of her personal genome will be finished before the end of the night-her replacement genes will be ready for us early tomorrow. It's getting late, Christopher, and Darlene really could use some sleep."

"Will you be all right, sweetheart?" I asked Darlene one more time.

"I'll be fine, honey . . . now go. I'm not afraid. I'll see you in a couple of days."

I kissed her goodbye and Billy followed me out. As we walked outside the main doors, Billy pointed and said, "That's my car over there."

"Very nice," I said, "What is it?"

"It's the new *Terra Hydra*; no gasoline, just hydrogen and oxygen. It doesn't have any exhaust pipes like the old cars. Instead, it has small drip ducts for the water it expels."

"Hum . . . how fast does it go?"

"She tops out at 180 mph."

"Interesting, maybe I'll buy one. Are you going to your store, Billy? I know its open twenty-four-seven."

"Nah, it's too late. I have to go home and get some sleep if I want to get there early in the morning-I'm usually there by six. But why don't you stop by on your way home and see what I've done with the place?"

"I think I'll do that. Hold it. Let's exchange phone numbers before we leave."

We followed each other until we reached the city and then Billy went one way and I went another. By now it was well past midnight, but I decided to visit Billy's store anyway. As I remembered, it had a big parking lot and it had plenty of cars parked in it. I saw a spot somewhere near the middle and proceed to park. I opened my door and began to step out when I felt a hard thump on the back of my head.

"Ouch! Son of a Bitch . . ." was all I could say before I turned around and saw a punk pointing a .22 and demanding money.

"Give me your wallet, motherfucker!" he almost whispered.

"I'll give you nothing you little bastard . . ." were the last words I heard.

I'm not sure if it was a dream or something for real. I remembered seeing my body first; and then the instruments and the top of three heads until Scott left the room, but that vanished quickly. I lost all sense of time and what I was doing. Nothing earthly seemed important anymore-not even Darlene.

It was the strangest and best feeling I've ever felt before. I was in need or want of absolutely nothing and still feeling a sense of wholeness. But then suddenly, I took great pleasures in the strange sensations as I moved very fast away, spinning through a tunnel of gray haze that sparkled with beams of warmth. And as the speed dramatically increased, I began to see a bright, bluish light, and the closer I got, the more urge I felt to be there. There was a faint echo of sounds calling me; I couldn't hear what they were saying but as I began to feel closer, a single force beckoned to me a dream with an all consuming, mysterious bliss; and just as I reached out to meld with it, I heard, "He's coming out of it."

When I opened my eyes I saw fuzzy images of Scott and a couple of other doctors standing at my side.

"How are you feeling, Christopher?" Scott said.

"Okay, I guess-just a little pain in my chest. Where am I?"

"You're in the hospital . . . and you're doing just fine." he replied.

"What happened to me?"

"May I be alone with this man for a few minutes?" Scott said to the others. After they left the room, Scott began to speak. "You were shot, Christopher-right in the heart."

"Who? Why? I don't remember."

"It was a mugger," Scott answered, "and he ran away into the night. You were shot with a .22 handgun. Tell me, Sandman-what does death feel like?"

"I died?" I asked Scott in disbelief.

"For a very short period of time-maybe four or five minutes, but you're alive and well now."

"Yes . . . I think I remember now. It was a very strange feeling-very hard to describe."

"You were lucky, Christopher. The bullet missed your ventricles and your pulmonary artery. It lodged in your right atrium as it passed through your superior vena cava. And the short time it took to Medivac you to the hospital helped saved your life. If it would've hit your aorta, we probably wouldn't be talking to each other right now."

"How did you find me?"

"GPS with an alarm on your vitals," Scott said nonchalantly.

"I thought you took those out?" I said somewhat angrily as I sat up in bed.

"Just the intrusive ones, Christopher, this chip is for your health and well being. And as you saw, it saved your live. I was at the hospital before you arrived, and I was the doctor who took the bullet out and glued you up.

You were bleeding quite badly internally and your Pulse Ox was very low. You survived by minutes, Christopher; you're either lucky or blessed-or both."

"*You* operated on me?"

"Christopher, I'm your primary physician. I'm licensed to perform many types of surgeries. I'm your neurosurgeon, your cardiologist, your pulmonary specialist, osteologist, urologist, and even your proctologist, just to name a few. If anybody does anything to you, it'll be me or through my supervision and approval. As long as I'm alive, your alive; Es aquel comprendido?"

"Yeah, I understand," I acknowledged, and then had to ask, "How long have I been here, Scott?"

"Just since yesterday night. I want you to stay another night, and then you should be good to go by tomorrow. Your car is at the Society-no damage. The thug shot you for nothing-there's still a good sum of money in your wallet-$800 or so if I counted right," Scott said, as he studied my medical chart while his voice faded near the end of his sentence.

"You counted right," I said irately. "Damn . . . talk about what goes around, comes . . ." I said to myself when Scott interrupted.

"There's a guard outside your door. I'll have Billy come and pick you up tomorrow, Christopher; I've signed the release form for noon. Billy will bring you back to the Society. By the way . . . Darlene's doing very well; I'm sure you'll want to see her."

When Scott left the room, the other two doctors walked back in and began a conversation and an examination in medical jargon, which they didn't think I understood, and so I pretended I didn't. They were impressed by Scott's brilliant handy work, and the rate at which I was healing. Only when they left scratching their heads, did I smile and relax in bed-and that was the last I ever saw of them again. I spent the rest of the day and most of the night watching TV until eventually I fell asleep.

"Time to wake up, Christopher," I heard a nurse say.

"What time is it?"

"Almost time to go home," she said as she gave me some pills and unhooked a tube. "There'll be a gentleman here at twelve o'clock to pick you up. There you go . . . my job is done," she whispered with a little grunt as she cleaned my arm with a cotton swab, and then continued to say, "There's a new set of cloths in the closet that Dr. Sellers brought you yesterday. You have about forty-five minutes to clean up before you go, and I'll be back with a wheelchair when you're ready to leave.

As soon as the nurse left, I hopped out the bed to brush my teeth, and then shaved, shit and showered, respectively. I toweled off, combed my hair and dressed myself with the new, medium-blue suit that Scott had left. Although I wasn't accustomed to wearing them, he also left a pair of silk, mid-size briefs for men. After putting them on, to my delight, I decided to go silk instead of cotton. Only moments after dressing myself, the door opened.

"Hello, Sandman, how are you feeling?" Billy asked somewhat embarrassed. "I'm sorry this happened, especially at my store-this has never happened before. If it helps any, I have at any one time, four surveillance security guards watching the store night and day.

"You don't have to apologize, Billy, "I said honestly, "the world is full of thugs, punks and killers. Besides, I feel as good as I ever did after I met Scott."

"Guess what?" Billy said.

"I give up," I said with a finger to my lip.

"I'll be driving you to the Society in my *Terra Hydra*. I think you'll enjoy it."

"We'll see; but it sounds like some goofy contraption out the twentieth century made in France or Germany."

About the time I was finishing my sentence, the nurse wheeled in a wheelchair for me to sit in while we commenced our leave from the hospital. I followed along with the tradition and planted my ass in the chair. And as I was doing so, I saw a small grin from the corner of Billy's mouth as the nurse started to wheel me out.

"What's so funny, Billy?"

"It's really not a matter of laughter, Sandman, but a matter of irony," Billy said.

"Yes . . . I see, Billy. You find it rather ironic for me to be sitting in a wheelchair when at one time you were committed to it for life. Trading places-how does it feel to have the shoe on the other foot. You're right, Billy, it isn't very funny."

When we made it outside, I spotted Billy's car with the name, *Terra Hydra* parked in the patients' pickup lane. To my surprise, it looked very ultramodern. I got out of the wheelchair, thanked the nurse and said good-bye. I strode slowly around the car, touching and looking. The finish was dull and gray and the interior looked like a spaceship. The tires were whitish and shinny.

"What's the deal with the wheels, Billy?"

"Cool-aren't they?" he said rhetorically.

"This is something new to me; what are they made of?"

"Some new type of rubber. They'll last for practically forever. Let me think . . . oh yeah; Scott told me they call them: B-T's.

"What the hell does that mean?"

"Okay . . . let me think . . . okay, he said they were made of something called: buckyballs dosed with titany terafluorene-or something like that," Billy said, mumbling at the end of his sentence.

"Fullerenes doped with titanium tetrafluoride . . . interesting. You know, chemists and material scientists are still having a field day with those little things."

"You know what they are?" Billy asked me quizzically.

"Organic molecules. Hollow clusters of sixty carbon atoms shaped like a soccer ball. Sixty atoms is the most common form found in a crystallized, face centered cubic arrangement of carbon atoms bound together with covalent bonds. However, the fluorine atoms in the titanium tetrafluoride are bound to the carbon atoms with ionic bonds. This is what gives your tires that whitish, shinny affect. How do they grip the road?"

"Like tiger claws. Even when I punch the gas, the wheels don't spin-it lays you back and smooth in your seat."

"Well, Billy, let's take a ride to the Society," I said.

"The Society?" He answered.

"Yes damn it; the Society. What do you think you're here for?" I explained.

"Oh, okay. Hop in and I'll drive. You're going to love this car."

"Just don't kill us," I said, "I don't like the feeling of death."

Billy pushed a button on a small device in his hand, and up went dove-wing doors on his car. When I sat in the seat, I could feel it adjusting to conform around me. It snuggled gently around my torso, to hold my position in place, while it left my arms and legs free to move about-essentially, it replaced the need for seat belts. The dashboard was completely digital and computerized, and it slanted downward so Billy could reach the recessed, butterfly shape of the steering wheel. He then pushed a blue-lit popup button beside the steering wheel, and when it turned green, we were off and on our way.

"Woo there, Billy boy; slow down now," I exclaimed as the seat stretched and leaned back to take up the slack from the G force. "What's your hurry?"

"I just thought you would enjoy the feel of the car."

"I do, but watch out for the traffic. Didn't you see the poor bastard you just ran up on the curb?"

"Ah . . . he's okay," Billy said as he looked into a rear view panel display.

"Damn, this is a smooth ride, and quiet too. Is there any internal combustion in this thing?"

"Behind the barrier in the rear of the car, there are tanks of compressed gases of oxygen and hydrogen. When they combine, they make water and heat. Scott told me that the water is exhausted and the heat is converted into kinetic energy of motion. There's also an atomic battery in there with some kind of turbine-don't ask me any more, because I really don't understand it."

"Hum . . . enthalpy, heats of formation; that's an exothermic reaction on the order of about 68,000 calories per mole, or approximately 270 Btu's from the moment you pushed the blue start button. And within fractions of a second, the energy exponentially increases until there's a shut off limit. Goddamn! I'm going to buy me one of these sons of bitches! When you come to a straight-away, give it some juice-I want to feel this thing fly."

And fly it did. Billy took his car up the mountainous roads going at speeds I wouldn't dare to try. It hugged the turns very tightly as my seat pivoted slightly to compensate for the centripetal force. It took us less than an hour to reach the summit and then we waited for a few moments until the outside gates parted. Slowly they opened to let us in. As usual, there were soldiers patrolling about, carrying their deadly weapons and walking the grounds. They didn't even look at us. We reached the front main doors and I watched Billy as he slid his card in the slot that made the doors open-like Frankenstein's castle-something I've never thought of before.

When we reached the midpoint of the interior lobby, we saw Sergeant Cooper standing and waiting.

"Hi, Sergeant Cooper, we're here to see Scott," Billy said in a happy voice.

Sergeant Cooper said nothing, but simply picked up a phone and pushed a button. "They're here," he said, and then ended, "Yes sir."

"Well, what's the tittle-tattle?" I asked Sergeant Cooper.

"Go to level five," he replied.

I looked at Billy and without speaking; I silently mouthed the words, "level five?" I knew that Billy read my lips, but level five didn't mean anything to him. We walked to the elevator and just before I pushed the button, I withdrew my arm and said, "Take us to level five, Billy."

"Okay," he said, and pushed the button, but nothing happened. "Damn thing is broken," he concluded.

"Wait a moment, Billy, let me try." I pushed the elevator button and it lit bright green as it read my fingerprint and blood type.

"Hum . . . works for you," Billy said with a grin. But on the ride up, I could only wonder what would be on the other side of that door, and then it opened.

"Plush . . . very plush. I'm impressed, Scott," I said to him while he sat in a chair facing us. "Who does your decorating?—The enhanced fairies of *Yorkshire Superfluities*?"

"Don't knock it; I fine it very comfortable, besides, it beats that two room apartment, or should I say dump, that you lived in on Steward and Browne streets."

"Touché," I said. "So why am I here instead of home? I haven't been home since . . ."

"Follow me," he interrupted.

Billy and I followed him to a private elevator that was shaped like a hexagon that I had never seen before. He said it works using pneumatics, and the ride was soft and descending. When the elevator door opened, we found ourselves in a corridor; again, a place I've never seen before. Sergeant Cooper was standing and waiting in the strange passageway.

"Sergeant Cooper," Scott commanded, "Please follow us."

We walked on down the ragged hallway. Scott took the lead while I followed, and Billy followed me. Sergeant Cooper took up the rear. Weirdly, the hallway didn't look like it belonged. After a few minutes, Scott stopped and turned to a heavy steel, black door while Sergeant Cooper unlocked it with a primitive key. It creaked when it opened. Sergeant Cooper stood outside while Billy and I followed Scott. To Billy's, and my surprise, there were two men chained face down on the dirty, concrete floor. Their arms, hands and legs were bound with shackles and their mouths were sealed with tape. They were filthy, ragged and murmuring frightened muffled hums. Scott kicked one of the men in his side and said, "Sandman, this man's name is Chester.—Turn your head around and look at me when I speak!" Scott yelled to the man furiously. "And Billy," Scott said with a pause, "this guy here is named, Spike," he ended his sentence this time with a phony inflection of concern.

Spike looked familiar to me, I thought to myself. Was that the same punk . . . ?"

"Sandman . . . it took at little time to find Chester, but we caught the bastard as you can see. This is the guy who shot you in Billy's parking lot-he's the one who put a bullet in your heart. He likes to play with guns and has itchy fingers-don't you-you motherfucker! And Billy . . . little Spike here was easy to catch-he's a leader of a gang and proud of it, or so he reckons. Do you know what he did to you?"

"What?"—Billy asked, but probably knew.

"He's the one who killed your wife and shot you in you spine!"

Never in my life since I've known Scott, have I've ever seen his eyes beam with such evil.

"Scott?" I asked, "What in the hell are you doing?"

"Revenge, my friend . . . revenge. Their lives aren't worth a plug cent. They're dead!" he yelled with spits of trails.

"But, Scott?" Billy started to say.

"Not another word, Billy. This isn't the first, nor will it be the last, that men have died who have crossed my path and choose to try and deprive me of my way. Billy?—I'm going to ask you to do something that's very callous and mean; but in my eyes, very justified. Sandman . . . give Billy your maser."

"My gun?" I said as I reached in my sport coat.

"Set it on kill," Scott replied, and which I did. "Now give it to Billy," Scott commanded, and which I did.

Scott bent down, and ripped the shirt up on Spike's back. He then pressed his finger hard on his lower spine. Spike started to scream behind his taped mouth.

"Do you see where I'm pointing, Billy?" Scott said calmly.

"Uh hum," Billy mumbled, and then looked at the gun.

"That little vertebrae is what we call the thoracic T12. Any damage to that little puppy and we're talking about paralysis and the lost of feeling near his groin-but not to worry; it won't kill him. Now . . . as I remove my finger, can you see the red spot I made?"

"Yeah." Billy simply said.

"That's where I want you to shoot him. But, before you do, let's remove the tape from his mouth to see what he has to say," Scott said, and then tore the tape from his face.

"Please, Billy . . . don't do it! I'm sorry for what I did . . . I didn't mean to kill your old lady . . . I didn't mean to shot you in the back. I'm sorry! Don't do it! For God's sake, please . . . ieeehhh-oh, that hurts . . . ouch, oh . . ." Spike yelled, mumbled and started to cry.

"Take that you bastard! Yeah-it hurts real bad—doesn't it?" Billy shouted with blood-speckled breath, and then giggled nervously.

I simply looked on in disbelief, while Scott looked at Billy proudly as he re-taped Spike's mouth.

After the tape was on, Chester grew wild as he flopped on his belly like a fish in a boat. He screeched behind his swathe.

"Billy?" Scott said above the noise, "Give Sandman the maser."

Billy handed me the gun. "Ahhh, come on now Scott; I'm not sure if I want to do anything," I said, but yet I did.

"This son of a bitch shot you in the heart-and for nothing!" Scott bellowed to me. "Let's see what he has to say," Scott said as he ripped the tape away.

"Don't shoot me in the heart. Please don't? Don't do it. I'm sorry, I'm sorry! All I wanted was your money . . . I panicked! God, you gotta believe me . . . I really . . ."

I spoke up in anger, "You know, you rotten lowlife; you killed me, Chester. I died. I was actually dead. Fortunately, I was revived; I survived. You wanna know what it feels like to be dead, old Chester, my unfortunate friend?"

"Please don't kill me . . . please don't," Chester begged.

"He's not going to," Scott replied.

"Then what am I going to do?" I asked Scott.

Scott looked at me with a strange grin, and then stepped on Chester's hand. Scott mashed down on the back of his hand until his fingers were spread apart.

"Sandman?" he said, "Shoot those digits off one by one-and take your time."

Billy looked at me convincingly; nodding his head and wringing his hands, and then I looked back at Scott.

"One at a time-let him feel some pain," he said.

And without a sound, a finger flew off, hit a wall, and then bounced in front of his face.

"Ahhhh!" he screamed.

"There's more," Scott composedly said.

And so off they came; one by one while the pool of blood submerged his paw. He yelled and screamed and begged until the five digits were gone. And then Scott stepped on his other hand and said, "We're not quite done."

I never realized how much blood a man's hand could bleed. But it wasn't long before the dirty deed was done, and then I felt the revenge. Sweet.

"Come on, boys, let's clean up and get a bite to eat," Scott said as he walked to leave the room. As we followed him out, I was surprised to see Sergeant Cooper standing at ease beside the door. Scott looked at him, and then the chamber of horror and said, "Let them bleed and suffer for awhile, and then get rid of them."

"Yes Sir!" he barked with a salute.

Billy and I followed Scott back to the hexagon elevator. He dropped us off on lever-four, and then told us to clean and change our clothes. "I'll be down in about an hour," he finished saying while we watched a seamless wall seal.

"Never knew that was there," I told Billy as I started toward the bathroom, while Billy headed for the bedroom. After a moment I heard him shout so I could hear, "There's two new suits laid out for us in here." I took my shower first.

In under an hour we were both sitting in the living room, sipping on a drink and waiting for Scott. We made small talk about Billy's store, consciously avoiding the reality of an hour before.

"How are we doing, gentlemen?" Scott's voice was heard as he entered the room seemingly from no were.

"Fine," we said in an awkward and unison way.

"Good . . . that's good. Let's step in the dining room for a bite to eat-shall we?"

CHAPTER NINETEEN

We sat down at the dining table exchanging glances and looks, but no one spoke until Henry the chef walked in. "Good afternoon Sirs; what's your pleasure?" Henry asked as he placed ice water, coffee and condiments on the table.

Not having seen good old chef Henry for a while, I spoke up and said with a smile, "How have you been, Henry?

He smiled in the corner of his mouth and answered unemotionally, "Fine thank you, Mr. Lance."

"Henry?" Scott asked, "May I have a dish of fresh caviar and Scottish smoked salmon with a glass of oak-aged Chardonnay?"

"Yes sir," Henry answered and then looked at Billy. "And what may I get you, Mr. Richers?"

"I'll have some good old fish and chips, and a beer."

"Very well, sir . . . and you, Mr. Lance?—What would you like?"

"May I get a steak filet and a baked potato with butter and sour cream?"

"Of course . . . and how would you like your filet?"

"Pink in the middle," I said.

"And what would care to drink?"

"I'll just stick with coffee, thank you."

"Very well," he said, and then disappeared through the double swinging doors.

"Well, boys . . . what's your feelings?" Scott unexpectedly said.

"About what?" Billy answered in a pretending way.

"Sandman?—What are your feelings?"

I sat for few minutes thinking, trying to justify ourselves, when Scott began a dialogue about the Holy Bible and the Old Testament.

"You, know . . . some people interpret the passage in the Scripture of Exodus 21:24 as a law of God's, and not of man's. I know the passage well, and I quote: "If men fight and hurt a pregnant woman so that she gives birth prematurely, and yet no harm follows, he shall be surely fined as much as the woman's husband demands and the judges allow, but if any harm follows, then you must take life for life, eye for eye, tooth for tooth, hand for hand, foot for foot, burning for burning, wound for wound, and bruise for bruise.

It contains a statement of the principle of fairness and a limit on retaliation. In Latin language, it's called: "lex tallionis", or the law of retaliation. Personally, I'm convinced that this is man's law, and not the law of God. So now, I asked you both-were we righteous in our actions, or are we sinners and condemned to hell? What do you think? Do we have the right to take the lives of renowned, and proven murderers? People who do the most ultimate act of sin of taking another human's life. I believe we do, or may God have mercy on our souls. I truly believe, that if mankind is given freewill and absolute, fair justice, than we are also capable of making our own laws too. Hell-humans have been executing their criminal citizens for eons."

After Scott finished his last sentence, Henry walked in, served our meals, and then left quietly and promptly. Scott began another monologue.

"The world is far off better without killers. First, there're useless to human society simply because they have killed; society doesn't need them, nor do they want them.

Second, corporal punishment is a proven and safe method to rid evil citizens from a populous striving to coexist in harmony, and sharing peace and more prosperity with every new generation. Thirdly, I have no intentions of geneticating human killers. End of story-do you guys have anything you want to add to want I've just said?"

"I think you've covered your convictions very well, and I support your position," I truly said.

"What Sandman said," Billy replied with serious eyes and with no repose.

I looked down at the table and was surprised to see that I had eaten most of my plate, even though I was unaware that I was even eating. I was listening too deeply.

"So what's going to happen to Chester and Spike?" Billy asked honestly, ending his next sentence with another question. "We didn't kill them-we just be crippled them-didn't we?"

"Wrong, Billy," I interjected, "they're dead men."

"Dead men?" Billy questioned while turning to Scott for his response.

"Sandman is correct, Billy," Scott replied, "They've seen faces and heard voices; I can't take the chance of letting them live. By tomorrow . . . it'll be as if they had never existed. All gone."

"Scott?" I said almost interrupting, "I don't care about those two miserable, forgotten bastards anymore-but I would be terribly grateful if I might be able to see Darlene again."

"Sandman?—You'll see her soon . . . tonight; and you both may go home together . . . tonight, if you wish. You too, Billy," Scott said easily and with an assurance of certainty. I took great comfort when I heard those words.

"When can I see her?" I asked Scott rather hastily.

"As I said, Christopher . . . soon. But first," Scott said as he got up and walked away from the dinning room, and which we did also to follow him, "I want to show you something Sandman. Billy?—You're welcome to leave now . . . I'll be in touch with you."

"Great, okay. Well . . . I'll see you gentlemen later," Billy said, and then added, "Give me a call soon, Sandman, and we'll have lunch."

"Will do, Billy. Take care."

After Billy stepped into the elevator, Scott began to speak. "We're going to take a little ride on that elevator too? There's something you need to see. And then, of course, you'll meet with your lady, Darlene."

"Just point the way," I said.

Scott and I walked to the elevator and then asked me to push two buttons in quick succession labeled: sub-level five-zero; somewhere I've never been before. The buttons lit bright green and the elevator door parted. We stepped in and Scott began to talk to me.

"You now have clearance to go anywhere in this complex, Sandman. And from now on you rank one under me-and you should know by now, no one ranks higher than I. In the coming years, you'll learn much more about this place . . . and the clandestine officials and agencies who we choose to cooperate."

"I don't know the first thing about this place," I told Scott with an element of bewilderment and uncertainty.

"In time you will, Christopher . . . in time you'll learn."

By now the elevator had stopped and the door was wide-open waiting for a decision for one or both of us to leave. The door closed after we stepped out and into a long, narrow and tall foyer. He then ushered me into a room and pointed at a huge, smoothly constructed obstruction. I turned to look.

"Here's what I want to show you, Christopher," he said. And with a wave of the palm of his hand over some iridescent device, the obstruction turned from solid to clear.

For the slightest moment, my breathing stopped, and then I gasped in awe. "My God . . ."

"Here is the beginning, of a new world and a new race: Homo genesis sapiens," Scott said with inflections of pride and overtones of destiny. "Take it all in, Christopher; magnificent-isn't it?"

From edge to edge, built in the mountain's interior, leaned a curved and slanted wall at lease twenty stories tall. At regular spaced intervals, with bright white lights, were attached oblong boxes that looked like coffins-thousands of them.

"Scott! This is massive! It looks like science fiction-are these coffins for the old race?" I asked him as I glanced around in amazement of the enormous task. "What am I looking at?"

Scott began to chuckle, with amusement unknown to me. "These are not coffins, Sandman, they're *genpods*."

After Scott said that word, it all became clear to me. "You've modernized the procedure-haven't you? And these things can now do what took you a month to do to me."

"Exactly. However, what used to take a month, can now be done in hours," he said affectedly. "Look at the computer we're assembling in the recessed center of the floor."

"That's a computer? I've never seen such a huge strange looking thing."

"Those were my sentiments as well, when it was first proposed to me a month ago. Its components' combined weight is approximately 80 tons. The computer is so powerful; it can sequence and identify every single atom in a human genome in less then a second. Its speed has been calculated to perform 100 giga-trillion teraflops. Can you even begin to grasp the enormous and monstrous pre-eminence that this thing has?

"Are you serious! Are you telling me the truth?"

"Hard to believe-isn't it?"

"You're goddamn right! But who? How? Where was this conceived and built?"

"Now we get into politics," Scott simply said, "and I'd rather not go there right now, Sandman-this is not the time, nor the place."

"That's fine with me, Scott, now when can I see Darlene?"

"She's waiting for you on the forth level-let's go."

When we were in the elevator, Scott informed me that we were indeed, going to create a new race, and then the door parted. I walked hasty to the living room and saw Darlene standing with her back toward me. I stopped and looked. Gradually, she turned around and looked at me without saying anything.

She was wearing a silk white gown, with golden lace that wrapped around her neck and extended down and around her waist. She had a head full of long blonde hair, which was styled with soft waves and modest curls. Her face was pampered, with beautiful makeup, which enhanced her pretty green eyes and soft pink lips. I had almost forgotten how lovely she was. Slowly, she began to smile.

"My God, Darlene, you look absolutely breathtaking. I don't know what else to say."

"Say you still love me," she said in a gentle voice.

"I love you," I told her as I stepped forward to hug her closely, and then kissed her.

"Oh, ouch . . ." Darlene moaned as she abruptly broke away. "You're hugging me too tightly, Christopher; I'm sore all over."

"I'm sorry, sweetheart, I didn't know."

"Of course you didn't, my handsome fiancé; kiss me again."

After our gentle kiss, I noticed that Nancy had entered the room.

"Oh, I didn't see you come in, Nancy, I was kind of preoccupied."

"Understandably so, Romeo," she answered with a smile

"Why does Darlene feel so sore?"

"It'll be gone by tomorrow," she said. "We have a new procedure now; it takes only days instead of weeks. She's just kind of bruised inside because all the injections happened in quick repetitions, unlike when we geneticated you, Christopher. As I already have said, soon we'll be able to geneticate in hours."

"Ah . . . yes, I know. I saw the room," I said as Scott looked at me.

"Oh. Okay . . . I see," Nancy replied while looking at Scott.

"When can we leave?" Darlene suddenly asked.

Nancy replied, "You're free to leave now if you wish," she said, then added, "and don't forget to take your medication, Darlene."

"I won't."

"Hey . . . I forgot-where's my car?" I asked Scott.

"It's parked by the front; here's your keypad," he said with his hand outstretched. "I'll be getting back to you soon, Christopher."

On our way home, I drove rather slowly as Darlene and I talked.

"Did you feel any pain?"

"None. It just felt like I slept for a couple of days. I dreamt about you."

"You did? What about?"

"You were living back in that small apartment again and Vern was still alive. All I remember was little bits and pieces and now everything seems smudgy and vague. But you and I were seeing each other without him knowing about it. I dreamt you were writing a book using long hand, and I was embarrassed to tell you that the book was really stupid. I know your not stupid, honey," she said to me and placed her hand on my lap.

"What else?" I asked.

"Just crazy stuff and other things I hardly remember," she said as she laid her head back in the car seat and closed her eyes.

I didn't ask her any more questions, and for a while, silence was all to be heard. As we slowly approached the city below, Darlene suddenly lifted her head up. She rubbed her eyes with her fingers before she spoke.

"Where are we going?" she asked.

"To take you home, sweetheart . . . unless there's someplace you'd rather go?"

"There is."

"Really? Where would you like me to take you?"

"I want to go home with you."

"Oh. That was a pleasurable surprise. You had me worried for a moment. I'd love for you to come home with me."

"How much longer before we get back," she asked me as the sun was slowly going down.

"Soon."

I turned down Salisbury Lane, which led me to my gated driveway. Finally, I parked my car at the front of the lighted entrance. It seemed a long time since I had slept at home, but when Darlene and I walked in, it only felt like yesterday.

"I almost forgot how lavish your home was, Christopher. It looks more attractive to me since the last time I was here." Darlene abruptly changed the subject. "I feel tired, Christopher, what time is it?"

"Quarter to ten, sweetheart; are you hungry?"

"No. I just want to lie down. Where do I sleep tonight?"

"With me, if that's all right with you?"

"That sounds splendid; I wouldn't want it any other way. Show me to your bedroom, my love."

My staircase was graceful and wide, with the steps covered with soft plush carpet. I carried her up to my master bedroom suite on the third floor.

"Oh . . . this is very nice, Christopher. Are those satin sheets?" she said as we entered my bedroom.

"Sure are. Have you ever slept in satin?"

"No."

"Good, I think you'll enjoy them."

"My I use your bathroom to freshen up?"

"Just around the corner, my dear."

While she was in the bathroom, doing whatever women do before they lie down for the night, I disrobed and crawled into bed. I was trying very hard to control my emotions, but as hard as I tried, I couldn't help but become aroused thinking of her sleeping with me. I waited anxiously hoping to make love to her again. She stepped into the bedroom and disrobed slowly in front of me. As I lay on my back, she saw the tent my penis made underneath the sheet.

"Uh, oh. What was I thinking," she said embarrassingly, "I feel too sore to make love to you tonight, honey. I'm sorry. I wasn't trying to tease you, Christopher—honest. Can't we just softly touch, hug and kiss. I've missed your affectionate caress and the sweet whispers in my ear. Can we just lie together tonight and enjoy some time of non-love making pleasure?"

Feeling disappointed and somewhat frustrated, I hid my selfish feelings and simply replied, "I wish nothing more than to make you feel safe, comfortable and relaxed. I love you and respect your wishes."

She gracefully laid down and slid between the satin sheets, which covered her naked body, and then rolled on her side to kiss me. I hugged her softly against my body and kissed her parted lips gently, and then whispered in her ear, how much I really loved her. We laid in bed whispering love words while taking turns caressing and kissing each other all over. Eventually, the non-love making disappeared gradually, and she laid on her stomach and fell silently asleep, while I softly stroked her silky hair, her neck and slender back. After one last somnolent kiss, I quietly turned away, closed my eyes and waited for sleep and dreams.

Slowly, I opened my slept-felt eyes, and heard the faint sound of a shower splashing.

I laid in bed and listened until the water stopped, and after a few minutes, Darlene walked in with a towel wrapped around her hair and completely naked. She looked at me and saw that I was awake.

"You know, honey, I'm going to need some clothes to wear if I'm going to be staying here."

"Just put on what you were wearing and we'll go to your apartment and get some clothes. What time is it?"

"It's about ten-thirty; are you hungry?"

"A little. Are you?"

"I'm starving."

"If you wish, we could see what's in the kitchen, or if you can wait for a little while longer, I'll clean up quickly and we can go to your place, change clothes, and then stop at a restaurant-how does that sound?"

"How about I just go down to the kitchen and fix us some breakfast? I'll try and have it done by the time you're finished in the bathroom."

"Whatever you want, sweetheart. Give me about twenty minutes or so and I'll be down to join you. What are you going to make?"

"How about some eggs, potatoes, bacon and toast?"

"Sounds like breakfast to me. I'll be down after I shower."

"Take your time, darling. This may take me a little longer then usual, because I'll have to search through your kitchen to find some utensils."

After I finished with my business in the bathroom, I dressed and met Darlene in the kitchen. I noticed, sitting down at the dining room table, a silver carafe of coffee nestled beside the cream and sugar decanters. Darlene had been busy.

"Smells good, sweetheart. Did you find what you needed?"

"Found everything, honey-have a seat and make yourself a cup of coffee."

"Thanks. You know what I'd like to do after we get your clothes?"

"What?" she answered.

"I'd like to buy another automobile-something that's not as attention-grabbing as my Jaguar XJR-9LM. Something that you could drive too."

"Sounds expensive. I hope you don't take this the wrong way, but how much money did you say you have?"

"A little over 300 billion now."

"Good Lord; that's an awful lot of money, and all you have to do is write?"

"That's about the jest of it."

"So what are you writing now," she said setting our breakfast plates on the table, and then sat down opposite to me.

"Science."

"What kind of science?"

"Well, at my last writing, I was working on some intriguing advanced mathematics."

"Are you writing a book?"

"More like sets of encyclopedias," I said.

"Just one set of an encyclopedia can take years to write and requires the collaboration of many professional and educated people."

"I'm not required of anyone's help. I can and prefer to work alone and besides, it takes less time."

"But how can you do that all by yourself?" she said as we began to eat.

"Well, at first I didn't think I could; even though I have an inquisitive nature. Come to think of it, when I first met Scott, I was always quizzing him about various things. Things I never really understood but wanted to know-and he always had an answer. Then came the genetication, the lump sum of money, the house and the car. He told me to start learning science. He said to start with the fundamentals and I'd understand the more complex science as time went by. And you want to know what?"

"What?" she said.

"He was right. I estimate my I.Q. right now at a little over 250, and I'm not sure if there's a limit; but there's a lot I still don't understand, and yet, I'm still learning and comprehending more advance information. You'll see for yourself-give it a couple of months and then try something that makes you feel good doing. Try painting."

"I hope we're doing the right thing," she said uneasily.

"Trust me-we are. By the way; how are you feeling?"

"I've never felt better, and my soreness has gone away."

After we were through eating, I showed her where to place the dirty dishes. "Just lay them in there."

"Is that a dishwasher? Don't I have to place them in a rack or something?"

"Nah, it's all automated. The damn thing eats the waste, washes everything, including the pots and pans squeaky clean, and then stacks them back where you found them-pretty neat-huh?"

"Very."

"Let's go get some of your clothes and buy a car; I'll be right back-I need to grab some cash."

I went downstairs to my office and into my safe. "Hum, one-hundred-thousand ought to do it," I said to myself, and then closed the safe and made my way up the stairs.

"Ya ready?" I asked with my keypad in hand.

"I'm right behind you."

We headed south on Steward Street toward the big, bad city. But we wouldn't be going near that corner where I used to live. We were going to a newly developed car sales business strip on Avirail Avenue. There were probably a dozen or more car dealers on the road.

"What kind of car shall we buy, Darlene? A sport car, or would a luxury sedan be more your style?"

"Most likely, something in between, and something that doesn't run on gasoline."

I made a right turn on Avirail Avenue and began scouting for car dealerships. "Let's try this one, or are the cars too out of the ordinary?" I asked.

"This might be a good place to start," Darlene remarked.

After our sixth dealership, Darlene finally spotted a car she liked. "I think this is the one, honey. Not too big, and not too small; and I just love the metallic, reddish-brown color. And look at the sticker on the window, it doesn't need gasoline-it runs on those new atomic batteries."

"What's the cost?" I casually asked.

"Oh my," Darlene said apprehensively, "the sticker says $88,000-that's too much."

"Where's a salesperson when you want to talk to one?" I said complaining, and then feeling somewhat foolish and asinine in front of Darlene. She appeared calm and contented.

"Here comes someone now," Darlene said as she looked in the direction of an overly happy man approaching.

"Name's Dave Sprocket, but you can call me easy on your pocket," Dave blurred out like an idiot. "So, I see you two have your eyes set on our new *Quantum Ununoctium 118*. She's all atomic you know."

"Yes, I'm familiar with the isotope. It was just two years ago, that the famous scientist, *Dr. Jacob Mckouski* achieved the process to obtain it in large quantities for commercial use," I responded.

"Hey . . . that's pretty good. All I know is it sure makes the *Quantum* go, go, go! Are you a scientist?" Dave asked curiously.

"I suppose you might say that, Dave. Does it matter?"

"Oh no," he said grinning again, "I've just never happen upon a real one before."

"How much?" I asked.

"Oh boy; that's a good question. First, I want you to notice that this thing's loaded. And you'll never have to stop for fuel. She has an expected life span of thirty years with a fifteen-year warranty. If you'll look here on the sticker, you'll see she's priced at $88,000. But I'll tell you what; I'm having a very good day today-I'll let her go for $86,000."

"Do you take cash?"

"Cash . . . hum . . . I think so. Follow me to my office if you would please, and we can start the old ball rolling," he said as he stepped in quick strides. We followed him to his office. "Please, have a seat while I make a quick call," he said as he pushed a button. "Hello, Linda? Is Fred in his office . . . okay, I'll wait," Dave said while holding up one finger, indicating this should only take a moment. "Hey, Fred . . . how are we doing today? Yeah . . . okay, yeah, right, that's good . . . hey listen, Fred, I have a quick question for you. Do we take cash?—Okay, yes, I see—up in finance, okay, no problem. I'll talk to you later-say what? Yeah, right, okay, alright, I can handle it from here, thanks Fred." Darlene and I just looked at each other wondering what the hell that was all about.

"Well?" I asked.

"Well what," Dave replied.

"Do you folks take cash?"

"I think so . . . I'm sorry, how could I ever forget-I can't recall your names."

"You never asked. My name is Christopher Lance and this is my fiancée, Darlene."

"My pleasure," Dave said, and then continued, "We do take cash, but the money will have to be scanned for authenticity up in financing-it should only take a few minutes. Now, here are some papers to fill out. If you need any help, I'll be right here and when you're finished and satisfied, just sign here, here and here." Dave sat back in his reclining chair with his hands clasped behind his head. He watched and waited.

"Okay, everything seems to be in order" I said, "just one last signature and everything will be legal. Do you folks deliver?"

"Well, that would cost extra . . ."

"Come on, Darlene, let's go somewhere else, I don't have time to play . . ."

Dave immediately interjected, "But on account that your paying cash, Mr. Lance," delivery is absolutely free. Where and when would you like it delivered?"

"Let's deliver it today. Darlene, give the man your address while I sign this page, and then we can head up to financing for cash authenticity and get out of this place."

After leaving the dealership, we decided to get a small bite to eat and a cup of coffee or two. We were searching for someplace different, but yet, we wanted a restaurant that had a feel for comfort and privacy. Suddenly, my car phone rang.

"Ciao, I'm listening."

"Christopher, I'd like to see you at five o'clock today," Scott's voice rang through.

"I'm with Darlene today; can it wait?"

"No, but Darlene may come if you wish . . . see you at five," and then the speakers were silent again.

"We still have time for some lunch and a cup of coffee. Hum . . . this place looks nice; would you like to stop here?"

"Why not? It may turn into a good place to dine out sometime," Darlene uttered unfeelingly.

The maître d' gave us a cozy corner cubical with engraved windows. They lightened our area with pleasant sunlight, and also gave us a nice view of some flowers, bushes and trees.

"I like this place, honey; what do you think?"

"I like it too, sweetheart. It has a comfortable feeling."

We ordered something light to eat and studied the beautiful scenery outside. The sunlight added warmth and serenity. It felt so goddamn great to have someone who loves you, and still respects the silence that's shared. I never felt any discomfort, as we said nothing, while we looked into each other's eyes. So silence never felt awkward, and neither did our conversations.

"Would you like to come with me to the Society, Darlene?" I asked nonchalantly.

"I'd rather not, Christopher, if that's okay with you? I don't care to go there again, unless I really have to."

"Would you like me to take you back to my place?"

"Could you take me back to my apartment? I'm kind of eager about the new car-I want to be there when it arrives?"

"Of course. But does that mean you won't be staying at my home tonight?" I asked her hesitantly.

"Heaven's no, honey. Just call me when you get home and I'll come back there with my new car-and probably with a suitcase or three."

"You had me a little worried, sweetheart. I thought you might have . . ."

"You think too hard, Christopher. I really want you tonight."

"What does that mean?" I had to ask.

"You know, silly . . . I need to make love with you again. I need you badly . . . and I feel so damn amorous. You're so affectionate and caring; I almost forgot how wonderful life can be. You know I love you deeply, thank you for such a wonderful time, honey."

I dropped Darlene off at her apartment, and then headed to the Society. I had a little trouble concentrating because I couldn't keep my mind off of Darlene. Whatever Scott wanted to tell me and/or show me made me feel a little impatient. I just wanted it to be over so I could be back with Darlene.

CHAPTER TWENTY

When I finally arrived, I saw the approaching gate to the building. Security let me pass and in the next few minutes, I found myself facing Scott. He had a device in his hand along with a brown envelope in the other.

"Here," Scott said as he handed me a cell phone.

"I always wondered why we weren't using these things."

"It not a cell phone, Sandman, it's a GPSSCD."

"Okay, I'm guessing that GPS means Global Positioning System, but was does the acronym, SCD mean?"

"It means: Satellite Communications Device. Basically, it's the same thing that's in your car.

"So what's so different about a SCD as compared to a cell phone?"

"We operate our communications in an entirely different frequency that's not allowed by the government for private commerce. We have six satellites, all in geosynchronous orbit with built-in redundancy. We operate in the higher frequencies of the spectrum: approximately, ten thousand terahertz, and that's close to the frequency of x-rays. Our communications are hidden within the spectrum-where no one else uses. Why only now-you might wonder, why I would give you that device? Because up until just recently, we didn't have the technical knowledge. We had to find a way to cancel the Casimir effect in our nano-technology, electro-mechanical relays. Do you know how it's done, Christopher?"

"I can't even begin to image."

"Neither could we, until a month ago. This breakthrough gave us the ability to make these SCD's as small as the one you're holding in your hand now.

No one can eavesdrop on our communications. Not even the government knows how to make one. Now, always carry it with you, as you do with your maser. By the way . . . show me your maser."

I pulled it out of the side of my sports coat and held it up to show him and said, "I never go anywhere without it."

"That's, good . . . but I'd like to get on with business. I asked you here for three reasons; one is that GPSSCD in your hand. The other two things I need to discuss with you are those genpods and a potential problem."

"You have my absolute attention and curiosity-I'm listening."

"Those genpods are near the stage of completion. There are two thousand and five hundred of them. We will soon be able to geneticate thousands of people per day, but the selection process will be confidential and a bit problematic."

"How so?"

"Well, for example, how would you decide to recruit the right people for genetication?"

"First, I suppose, I'd start with the people who work here, and then branch out to their relatives . . . but I'd have to think it through so it wouldn't be problematic, but still confidential."

"Well, Sandman . . . you're close, but there're other more important people to consider. Bradly is taking care of that."

"By the way, Scott, just how many people do work here?"

"Eight-hundred and sixty-four. And on any given day, about seven people leave to go back to their homes and personal lives. And on any given day, about seven people arrive back from their leave. On average, all our workers are arriving and departing within a four to six month cycle.

We have everything that's needed for all our personnel. For living, sleeping, eating and pleasure, and that's just to say the least. We also do research, materials development, design and manufacturing, fabrication and construction, recycling and a host of other processes. Where do you think all those genpods came from?"

"Obviously, from here," I replied.

"You're right. They were designed, built and installed in this complex. Very little material is transported to, or away from this place," Scott said proudly, crossing his arms against his chest.

"Now, I'm perplexed," I said, "Where do you get all your materials from, and what do you do with the waste?"

"Ah . . . the *Recycling Center*," he said with a pause, smiled, and then continued; "it's a complicated process of breaking down mass by using a combination of mechanical, chemical, electrical and nuclear systems. The waste goes into one end, and comes out at the other as separate and pure atomic elements. We even recycle the exhausted gases and the air that we breathe."

"Fascinating. What about food? How do you feed everyone?"

"We use hydroponics and livestock through breeding and cloning. We could survive on the moon. But now I'm starting to diverge. As I said earlier, we will soon be geneticating thousands of people per day, and as I also stated, the selection process will be confidential and a bit problematic."

"So how are you going to proceed?" I asked Scott sincerely.

"Everyone in this complex who wishes to be geneticated will be so. And everyone in this Society is aware of the confidentially and the consequences of breaking our silent code. The disclosure of this place is an automatic death sentence with one exception."

"And what is that exception?"

"They are allowed to tell only their wives, children, and one of their youngest and closest friends. But they will be bugged with an implant, which they will be aware of, so we may monitor what they say and to whom. Any unwanted acknowledgment of us to anyone other than those we approve would be their demise, including the unsolicited people they talk to. I'm hoping that this doesn't happen, but in all probability, it will, and this is the problematic aspect of our plan."

"How would you have those that break the code killed?"

"We have our assassins."

"Oh really?

"Don't ask me anymore questions in that regard, just be aware that we have them-end of story.

"Understood. When do you plan on starting the geneticating procedure?"

"Within a day or two."

"Were those the things you wanted to talk to me about?" I asked.

"No, there's one more thing, Mr. Sandman. Again, as I said earlier, there're other more important people to consider, which an assassin is taking care of as we speak."

"I'm listening," I said as Scott opened the brown envelope.

"Let me give you a short rundown of some of our more important clandestine representatives, Sandman. First, we have two operatives working in the United States Department of Defense for the Advanced Research Projects Agency. Here are their pictures. The man is known as agent, Alpha-one, and the woman agent is Alpha-two. I want you to remember their faces-all of them.—How's your memory?"

"I can recall conversations, written text, photos and videos to about 75% accuracy at this point, Scott, but it's improving as time goes by."

"Very good. Here's a file disk, and this is your password. Commit it to memory and install the disk on your computer. After installation, the disk will become corrupt; simply break it in half and burn it. Once it's on your computer, I want you to memorize all their faces, birth names, backgrounds, powers of duty and responsibilities, and then make sure this information is kept locked in a hidden and encrypted formatted file.

Let me show you a few. Here's a picture of agent, Beta-one. Bata-one is a senior official for the Department of State; he's very reliable and essential to us. These two agents: Gamma-one and Gamma-two work for the Department of the Treasury in the bureaus of the ATF and the IRS. And this woman is agent, Delta-one; she works in Customs Service.

Here we have agent, Epsilon-one; he helps us with oversea operations through the Department of Transportation.

This picture is agent, Zeta-one. Zeta-one works in the General Counsel of the Department of Energy. He's been very helpful to us many times in the past. As a matter of fact, he just recently helped supply us with the enormous amount of energy that'll be needed to run our new eighty-ton computer. And there's more-study them.

Now . . . you do realize that if you divulge any of this information, you'll breach our security and expose us to the world? And in doing so, many lives will be lost and the consequences will be formidable and unpredictable. My guess would be that internal, and then global conflicts would be the first two destructive waves. We pose a potential catastrophic danger to the entire world. Do you understand the serious gravity of this operational arrangement, Sandman?"

"Yes sir; I do indeed."

"Will you commit your complete alliance to me, Sandman?" Scott inquired.

"Absolutely, sir," I said in all honesty.

"Good. But I must sincerely warn you, Christopher . . . communication of anything that goes on here, to any unauthorized person or persons, may result in your death too. However, Christopher, we know that you have never broken this rule, except to Darlene, which I preferred to overlook. We simple have too much at stake. Besides, I want you to become an integral part of the Society."

"So what is this assassin doing as we speak now?" I asked.

"He's assassinating one of our agents who's breached our secret code and who's been talking about the Society to an unauthorized government employee. If our operations are leaked too soon, we'll never be able to complete our mission."

"And what is our mission again?" I asked for confirmation.

"To alter and change all the humans into a new species-like us."

"You mean the whole fucking world!"

"Well, Sandman, not all humans; just the ones who are appropriate. In the general populous, we'll be considering only those who are thirty years old or younger. Of course, there will be exceptions on a case-by-case basis. For instance, maybe a wife and husband who differ by a few years or two, or possibly someone who has an exceptionally high I.Q.-those kinds of things.

"Why are you telling me this, Scott?"

"Because, as I just said, you will become an integral part of the Society, and a very important person to the entire world."

"What do you mean by that?"

"You're very special, Christopher-more than you know. You, and only you, because of your genetics, will reach the summit of the entire human and nonhuman race. You're in for a wild ride, my friend, and don't ask me to explain it-you'll know when it comes. Now, take this computer disk home and do what I said."

"But . . . I'm not sure what you mean?"

"Say nothing more tonight; I'll be in touch-now go home to your future wife."

On the way home my thoughts drifted in a muddled and disorganized hodgepodge of possibilities. How in the hell did Scott really keep the Society hidden from our government and the rest of civilization? Does the threat of death really work? Was he paying them hush money too? Did he have a squad of assassins ready at a moments notice, or did he really have all his people aligned and committed to the projects that the Society had done and are still working on? I knew I wasn't going to tell anyone. Maybe that's how all the other people felt. Maybe being geneticated was their reward? I couldn't quiet understand how such a big and powerful organization could be kept secret for so long. And on it went, rambling possibilities and discontented contemplations. Finally, I let it go as I got closer to home. Darlene was now on my mind. I was becoming anxious to have her with me again, and as soon as I got home and walked through the door, I was on the telephone asking her if she was still coming over. Her answer was yes, and suddenly I felt a silly, giddy type of joy-like a schoolboy who just got his first kiss.

Darlene was at my gate by eight-thirty and the sun was resting low in the sky. I pushed the gate button to let her in, and she pulled up to the front door in her new Quantum Ununoctium 118. I went out to meet her.

"Hello, Sweetheart. I see they delivered your car-very nice. Did everything go well?"

"Everything went just fine, honey. Would you help me with some of my clothes and these boxes I brought along?"

"Of course. You go inside and I'll bring them in."

She told me that some of her displays and pictures were in the boxes and other personal belongings packed in a few suitcases. I carried the suitcases upstairs to the bedroom, while she instructed me where to put the boxes. Her hair was a little messy and her white jeans and blue top were a bit soiled.

"Would you like to clean up?"

"Eventually," she said, and then asked me a question. "Did I imagine it, or did you once tell me you have a swimming pool?"

"I believe I told you."

"Good, I want to go for a swim-will you join me?"

"Right now?" I asked.

"Yes, right now. Show me the way."

She followed me as we walked through some corridors that took us to the pool.

"Oh my gosh—this is absolutely delightful. How high is that glass ceiling?"

"Two stories. Would you like me to open them so we can swim under the stars-it's just about dark outside now?"

"Yes," she said, and then after a few moments when the ceiling began to open, she replied, "Oh, Christopher; this is absolutely heavenly!"

It didn't take long before she had stripped her clothes off and had her feet dangling in the water. "Perfect-it's like bath water!"

I took my clothes off and dove in and when my head popped out of the water, I caught a peripheral view of Darlene diving in. When she came up, she swam to me and gave me a kiss with a strong wet hug. I felt her naked skin against my naked body. And as much as I tried to suppress it, I couldn't. I had grown an erection.

"Oh, my goodness," she said playfully, "What do we have here?"

I ignored her question and starting to kiss her lips and breast while I gently stroked my penis on top of her vagina while at all times, avoiding any penetration. I wanted intercourse to come later, as I expected she did too. After a few moments, we casually split apart and began to splash each other and doing other playful things. We even raced each other across the pool and did a few jumps off the lower diving board. Eventually, we ended up side-by-side floating on our backs while looking at the stars.

"This is so nice, Christopher . . . I sure do love you."

"And I love you too, sweetheart. I've never felt such a deep devotion as I have with you. I can feel your soul immeasurable inside me, and from the songs that sing to my body, I can tell you feel the same way about me."

"Honey," she said as she rolled on her side and hugged me tightly, "take me upstairs so we can make love-I need to feel you inside me."

I carried her upstairs and on silken sheets, we made love more passionately then we ever did before. We meld together and embraced each other much more then I could ever remember. And the pleasure of ejaculating in her over and over was nothing I've ever experienced prior to this enchanted night.

She confessed to me the very same feeling that I had felt with her. She told me that she had many intense orgasms that were beyond any that she had had before.

She cried some soft tears, which dropped on my shoulder as she laid on top of me while I gently stroked her hair. After some tender spoken love murmurs, we peacefully drifted off to sleep and woke the next morning cuddled in a spooning position that lovers usually do.

"Christopher?" she said softly in a morning whisper, "I've never felt our love making so passionate and exhilarating as I did last night-did you feel it too?"

"Without question, Darlene, indeed I did. But I'm sure I know why."

"Tell me," she said as she rolled over to look me in the face.

"It's because you, and I, have been geneticated. Everything is heightened. We feel things humans can't. Just like me, all of your senses are becoming more acute. All of your senses, and than some, feel more pleasurable, while pain does just the opposite. As time goes by, you'll gain more strength in your body too. And your mind will start to understand things that you could never understand before. You'll never catch another cold or flu and diseases will never invade you. Cancer will never threaten your body, and you'll live as long as I will. Do you follow what I'm saying?"

"Yes, Christopher . . . I do. It's just so hard to believe."

"Trust me, sweetheart, in time you'll see its true."

Darlene perked up and began to smile, and then said, "Who wants to get cleaned up first; me or you?"

"Lady's first," I said.

While Darlene was in the bathroom, I rediscovered, as I looked out a window, my guesthouse that stood some distance away on the back of my property. But, because of my viewpoint, it was hard to see and also because it was partially obscured by those large pine trees. "How in the hell did I forget about that?" I said out loud to nobody.

I sat on the edge of the bed and thought for a while. We could have someone to cook our meals, clean our house and even change our bed. But I wasn't thinking rationally. Granted, our house was big, but cleaning it certainly isn't a full time chore. My house was pressure positive and it would take at least a year before any amount of dust would accumulate. The vacuum cleaner keeps the floors clean and polished, and it'll even do the stairs. Besides, I believed she would enjoy decorating and furnishing our new home with the things we each enjoy. A commerce provided by the Society-just like my food service, would launder all of our clothes and other things. However, I knew that Darlene still enjoyed doing her own cooking.

Although I had more than one bathroom, we took turns using the one in our bedroom because everything we needed was there and it was our favorite.

After Darlene came out of the bathroom smelling so fresh and good, she came down to the kitchen where I was savoring a cup of coffee.

"My turn to clean up," I said, "and afterward, I want to take a look inside my guesthouse."

"You have a guesthouse?" she asked.

I escorted her to the back entrance and opened the door and pointed. "There it is, surrounded by the large pine trees."

"Oh, yes, I see."

After forty minutes in the bathroom, I came downstairs with my keypad to see if it would unlock the guesthouse, and before I left, I asked Darlene if she would like to come too." Her answer was yes.

We walked down a cobblestone pathway enjoying the warn summer breeze, the smell of flowers and fresh air and the pre-afternoon sunshine. After a pleasant five minute stroll we were at the front door. I soon found that my keypad that lets me inside my house was not needed for the guesthouse-it had a simple doorknob. As I began to turn the knob I discovered that it was locked. With a strong grip and a steady twist, the doorknob broke free. I slowly swung the door open and stepped inside while Darlene followed behind.

The room was grim and dim with its windows being hidden by the large pine trees. It was dusty and had a smell of antiquity with very little furnishes. The air tasted old and stank. I did a slow 180-degree turn while I observed my surroundings. There was nothing out of the ordinary, just an old unkempt home. There was a second floor and I walked in the grayish fawn colored light to the staircase. All the while, Darlene was standing close to me.

"I don't like this place, Christopher . . . let's leave."

"After I take a look upstairs," I replied softly.

"This is a creepy guesthouse, Christopher, let's get out of here," she whispered as she followed me up the stairs.

There was a small hallway at the top and I began to walk to see what was up here. As I reached the top and took a few steps, I suddenly heard the click of a gun hammer cocking back and I knew that it was locked and loaded and aimed at my head. I made an extemporaneous effort to stand still.

"Slowly step back away from me and don't make any sudden moves, or I swear to God, I'll kill you."

I slowly backed away with Darlene behind me. "What do you what?" I asked the man.

"Just a chance-I knew I'd catch you soon-it was just a matter of time and now, here you are," he said with a sweaty forehead.

"A chance at what? I don't even know you, my friend," I said calmly.

"I know you don't know me; but I know you."

"And who am I?"

"If I'm right, and I'm sure I am . . . your name is Mr. Sandman," he said nervously with wet, matted down hair and still pointing the gun at me with an angry stare.

"What do you want?" I asked again.

"My family is dead . . . they're all dead! And it's because of you and some kind of conspiracy. My brother in law told me. He told me lots of things. I didn't believe him at first when he told me they were going to kill him, but now I do. He told me where to find you and seek his revenge-and now my revenge. You guys blew his fucking house up! His whole goddamn family was killed! And so was mine-you're some kind of genetic freak-and I know there's more of you." he finished in a tirade of grumbled words and spit.

He was pointing a .357 straight in my face. "Maybe you're right," I said trying to placate him, "but maybe he was wrong and there's no such thing." Darlene stood quietly behind me and I could feel her legs trembling. I knew I needed more time-time to distract him and then attack him. I knew my reflexes were much faster than his. "Where did you hear such a crazy idea? I'm not sure if you know what you're talking about."

He started to yell. "You had my sister's family killed, and then you killed my family! And now it's your turn to die!" he yelled as he fired his gun. Before the bullet was out of the barrel, I had pushed Darlene to the floor. And within a split second later, I had his hand in my hand, still holding the gun. But his hand was no longer attached to his arm as he fell to the floor. I quickly stomped on his throat pressing my foot down hard while I slowly applied more and more pressure. His chin and chest bent oddly toward each other as blood began gurgling from out of his mouth. In less then two seconds, this man was dead. The skin around his mouth was stretched in an odd way that gave him the appearance that he was smiling as he was dying.

When I laid the man's hand down on the floor, still holding the gun, Darlene lost it. "Oh my God! Oh my God! What in the hell just happened! He tried to kill us! You killed him-he's dying! I've got to get out of here . . . I've got to get out of here," she screamed as she ran down the steps and then out the front door to get back to our house.

I chased after her and found her face down on the dining room floor crying hysterically. I bent down to console her. "It's okay now, baby. There's nothing to worry about; I wouldn't have let him hurt you. This was a mistake-this shouldn't have happened. I promise you . . . never again. I'm sorry, sweetheart. I'm going to give Scott a call right now. I want to get to the bottom of this. Come on, sweetheart; let's go into the living room and I'll fix you a drink with a sedative. You'll feel better in a few."

Darlene reluctantly got up as I helped her, and then she staggered into the living room where I sat her gently in a chair. I poured her a drink and fetched her three *Valiums* hoping they would still work on her. After about five minutes, she seemed to relax and was talking more coherently.

"Honey," she said to me, "did that incident have something to do with the Society?"

"If I had to guess, I would say yes, I'm calling Scott right now," I said as I pushed a button on the phone.

"Hello, Christopher, what can I do for you?" Scott said joyfully over the phone.

"Could you have someone remove a dead man from my guesthouse," I said placidly.

"You caught him!" Scott perked up.

"Why didn't you fuckin' tell me; who was he?"

"Nobody special, but important enough to start trouble. He got away from our assassin. I'm sorry if we've caused you any problems. I really had no idea that he would show up at your place. Is everything all right?"

"I'm okay, but he frightened Darlene terribly. She was with me with when he tried to kill me. How did he know where I lived?"

"Most of our government agents know how to find all of us. And as I mentioned to you earlier, one of our agents started squeaking, and that's apparently how this man found you. I'm glad to hear that you're okay. It's really my fault; I should have cautioned you, forgive me, Sandman; and give my regards to Darlene and tell her I'm sorry. I'll send some men over to get rid of the body-you'll never know they were even there. Can I help you with anything else?" he said calmly.

"Yeah . . . send someone over to tear down that goddamn shed of a guesthouse thank you," and then the phone went dead.

"You know, sweetheart, sometimes I really don't understand that man. Sometimes he really burns my ass. He's a tough bird to figure out. I trust he knows what he's doing. How are you feeling-better I hope?"

"I'm okay. It's just that I've never seen something so violent. I've never seen anyone killed before, especially by you. But I pray that you're okay. If he had killed you, then he would have certainly killed me. I couldn't stand to see all the blood that was squirting around when you twisted his hand off his arm. It happened so fast.—Are we safe, honey . . . I mean, are we really going to be okay to sleep here tonight?"

"We're safe and we're going to be okay-it'll never happen again. I promise. Now, I'm going to change my shirt and wash this blood off me; will you be okay by yourself for a few minutes?"

"Yeah. I'm starting to calm down. I'll be okay.—Boy, that was really something. I need another drink."

"Would you fix me a scotch while you're at it, please, and I'll be back in a moment?"

I returned wearing a clean shirt and sat down next to Darlene while she handed me my glass of scotch. She appeared completely composed and even started a conversation. In a way, this surprised me, but after a few minutes we were talking pretty much normal again.

"Christopher, how did you know he was going to shoot at you?"

"I heard him pull the trigger."

"You could actually hear that sound-I didn't hear anything."

"In time you will. In time you'll learn a lot of things you could never perceive. As strange as this may sound, I can now filter out different tones and concentrate only on the ones I want to hear. I could literally hear that distinct sound as he squeezed that trigger."

Darlene seemed so relaxed that I'd swear the incident had never even happened. The Valiums must have taken effect.

"Is someone coming to get rid of the body?"

"It'll be gone before the day is over."

"Why did this guy try to kill you?" Darlene asked sincerely.

"He was caught in a breach and was trying to create a leak in what the Society was doing. That's an automatic death sentence."

"But Christopher, don't you believe that this information is destined to reach the public sooner or later?"

"Indeed I do. But by the time that happens, there'll be no turning back. The population will want to become geneticated because of the advantages. The ramifications, however, may be staggering."

And so we sat drinking and talking about the possibilities of what the Society was doing and what would happen to the world in general? We sat up talking late into the evening and finally retreated to the bedroom where the talking stopped and the sleeping began.

I awoke early the next morning while Darlene was still sleeping. Quietly, I put on my robe and took a small stroll to the guesthouse. Inside, everything looked a bit cleaner and when I walked upstairs it was no surprise to me to see that the dead man's body had vanished. I returned to the house and made myself a cup of coffee.

Darlene came down from the bedroom and caught me standing at the kitchen bar in a daze of thoughts.

"Is he gone?" she said as she rounded a corner, and then carefully poured herself a coffee.

She awoke me from my daze. "Yes, Darlene . . . he's gone."

"What do we do now?"

"I'm not sure. I think I'll check in with Scott and see what he has to say," I said as I walked away and into the living room to pick up the phone.

"Hello, Christopher, how are you this morning?" Scott said over the receiver.

"Okay, Scott. Listen . . . I need to ask you a question or two?"

"Let me see if I can guess, Sandman. You want to know what happens from here and what you and Darlene should be doing. Things kind of appear pointless at this time and you're feeling a bit confused. Am I right?"

"Yes . . . you're right. What's going on?"

"Don't concern yourself about the present; everything is going as planned. I want you to get back to your academics and Darlene can do whatever she wishes. I'll get back to you later."

"Hold up, Scott, don't hang up just yet. I'm beginning to lose my perspective on why I'm doing all this studying and writing-give me a clue because I can't remember."

"Okay, Sandman, this is how Karen explained it to me. When you were cloned, she noticed a discernable amount of activity in your temporal lobes, near the hippocampus in your brain. It's very active, and the area extends beyond normal human beings. It became very apparent to us that you could be re-engineered to have a strong desire for knowledge with the ability to comprehend it. We knew that you would become very clever and serious in science. When we re-engineered you, we exploited that part of your brain-we were quite sure that you would become genius-and now you are. I anticipate you will learn new science beyond what humankind already knows. This is what I want you to do, Christopher, follow you natural feelings; can you learn to live with that? Is this idea absurd, or might it be a noble way for one to spend one's life?"

"Okay, Scott, Karen and you were right; that's exactly what I want to do . . . I do feel an insatiable desire to learn and discover. I suppose I just needed some reassurance. I was becoming unsure. With everything that's been happening to me, I felt like I was losing my mind."

"Nonsense. If you'll recall, Christopher, this is one of the reasons why I invested a good sum of my wealth into your life, and to your future-our future. You were also the first in our audacious plan to evolve the human race, and it's just beginning. Now, spend your time as I asked, and spend some of it with Darlene. Take in every moment, but also be cautious. I'll get back to you later."

After the phone went dead, there was nothing left to do but follow his advice. I talked with Darlene and she agreed with Scott. After we talked, things became clearer, and I felt more focused. Darlene and I agreed to put that upsetting incident behind us and start out fresh again. Unexpectedly, we got on the subject of her apartment and her belongings.

"I'll just pay off the lease and then we'll be done with it," I told her.

"What about the rest of my things?" she said despondently.

"No problem," I answered, "We'll just hire some movers-problem solved."

By the end of the day, it was done. We had her bedroom set placed in another room, while she displayed all of her personal items around the house. We discarded her dining room set and other such things to the Salvation Army.

Most of her clothes were moved into our bedroom closet. She was officially moved in and we were both free from any unease. We were now living together and looking forward to marriage.

As time went on, I found myself spending more of my life down in my office and loving everything I studied, especially the subjects on mathematics. Gradually, as I progressed in the advanced areas in math, I realized that there were beautiful mathematical constructs and theories that seemingly had absolutely no counterpart in reality. I found this very intriguing. I was also subtly surprised at how math plays such an important facet in everything in life. Think of a subject, any subject, and you'll find mathematics. I was also absorbed in chemistry, astrophysics and propulsion systems. But the most intriguing aspect of reality, which I wanted to uncover and discover, was gravity.

I understood, in the most mathematical sense, the conflicts involved between quantum mechanics, gravity and general relativity and the search to merge the three.

But there was something very uneasy in the supposed solution with *superstring theory* that possesses eleven dimensions, membranes and parallel universes. The math was elegant but the ability to prove this GUT (Grand Unified Theory) was beyond reproach using the scientific method, and that was something that I just could not accept. I simply dismissed string theory as nothing more than an intriguing mathematical construct with no bearing on reality. I decided to tackle the elusive force of gravity and its mechanisms in my own way.

In the last couple weeks, Darlene spent most of her time reading, relaxing, sun tanning, swimming and shopping. She was on the vacation of her life and loving it. At the end of the second week, I popped the question and we were married in front of the justice of the peace with Billy as our witness. I have to admit, I was quite surprised at Billy's appearance when he met us at the court. He was well groomed and dressed very handsomely. His cologne smelled as fresh as a summer's evening breeze. Just one look at him and you knew he had money. After the wedding, the three of us went to an exclusive and fine restaurant that revolved at the top of the tallest building in the city. We had a few laughs as we ate and drank and talked about how everything was going so good. It was nice to see Billy again, but after we left, Darlene and I went directly back home to consummate our marriage.

We didn't go on a honeymoon but elected instead, to just stay home and play.

We went out to see movies, ate at nice restaurants, partied and danced at expensive and fancy nightclubs as well as shopping for things we needed or wanted. In essence, we spent the week having fun while doing whatever we desired.

After a week of celebration, things slowed down and became a bit more normal. I returned to my studies while Darlene spent much of her time painting. She enjoyed learning new techniques as she discovered her own creativity and I enjoyed reading, writing and learning more and more about science.

Life sure gets easy when you have money, unless you spend all your days worrying about losing it. I prefer not to worry, besides, I have my studies and Darlene to keep me busy. Never in my life have I felt so contented. And I've finished learning the entire subjects of mathematics that has been discovered and written to present. I've compiled them into an encyclopedia on my computer that can be read from start to finish. But of course, there are always new discoveries just around the corner and so this text will never really be finished.

CHAPTER TWENTY ONE

I've now devoted some of my time working on my encyclopedias of physics and chemistry.

However, I've also been fascinated by some particle physics in the contents of gravity. If my math is right, and I believe it is, then I've come to the conjecture that only neutral particles interact with gravitons, especially the neutron. I've drawn up some plans on a containment device, which looks like a flatten ellipsoid that allows me to create a variable neutron orb. Tomorrow, I'll download the plans to the Society and see if they can build my device. If it works correctly, I'll be able to vary the size of the neutron orb, which in turn should make the device heavier or lighter. If the experiment works, then I'm half way to establishing an anti-gravitational device. In my conjecture, I've established that the best emitter of anti-gravitons which should be anti-neutron particles. And if I substitute the variable containment device with anti-neutron orbs, I should be able to make the device float, canceling gravity and even levitating my device to higher heights. I'm also creating a new propulsion system, which works on the principle as the quark subparticle ultra-strong force, but with split quarts particles instead of elementary particles.

The spent energy I've calculated should be adequate to propel a spaceship faster than the speed of light. Doing some very advanced calculations on my computer, it appears that when faster then speed of light is achieved, space becomes one dimensional and travel becomes instantaneous and time stands still. I've been working steadfast for two days and now I'm wrapping up the evening early so I can spend some much needed time with my wife.

As I approached the top of the stairs, I saw Darlene lounging on the sofa watching TV. I laid down beside her.

"What are you watching, sweetheart?" I said as I wrapped my arm around her waist and kissed her cheek lightly.

"It's a documentary on volcanism and the cycle of life.

The volcanologists are predicting that the day the volcanoes quit erupting will be the same day when planet earth starts dying. It's really quite interesting. They're comparing earth to mars."

"Have they talked about Olympus Mons and the ancient fossils the astronauts found in the Valles Marineris canyon system?"

"Yeah."

Just as I was preparing to ask Darlene another question, the phone rang. As I got up from the sofa I mumbled to Darlene," This has to be Billy or Scott."

"Hello?"

"Hello, Christopher, this is Bradly; what time do you have?"

"Five to six."

"Watch the news at six o'clock and then get your ass up here as soon as you can-Scott wants to talk to you."

"What's up?" I asked mystified, "I haven't slept in two days."

"Just do as I said," Bradly answered before the phone went dead.

I stood for a moment with the receiver in my hand when Darlene spoke. "What's wrong, honey?"

"I'm not sure," I said as I hung up the receiver, "Scott wants me to watch the news at six and then immediately drive to the Society. Turn on the news channel for me; would you please, Darlene?"

I sat down in a chair and Darlene sat up in the sofa as we waited for the commercials to stop. Finally the news came on and the anchorman started to talk.

"Good afternoon, this is *World News Network Now* and I'm Peter Adams. We want to start with some groundbreaking news with a strange story. Three men and one woman walked into the Casper County police precinct and announced that they were cloned. Although there has been substantial research in this area, it is commonly known that there has never been a human cloned. However, these people have made incredible allegations about a hidden group of military scientists who have rearranged their molecular DNA to give them physical properties that are unlike normal human beings. And as bizarre as that may sound, two other individuals have claimed the same thing at two different county hospitals. Local authorities working on the case said they are investigating this strange matter and will report back when further details become available. And now we turn our attention to the precarious country's of Iran and North Korea . . ."

"That must be it, the word is out and Scott must have something up his sleeve," I told Darlene as I grabbed my coat and headed toward the door. "I'll be back whenever," I said and kissed Darlene goodbye.

"Please be careful," I heard her say as I closed the door.

I climbed in my Jaguar XJR-9LM and hit the cement. After I left the city streets, I floored the gas pedal and shifted rapidly through the gears. I topped out at 180 mph climbing up the desolate, black mountainous roads to reach the Society's front gate, and in less then an hour, I was there. I was half expecting a media show but everything appeared the way it always did. The gate opened to let me in as a guard stood vigilant by the side. I pulled up to the front door and went inside. Sergeant Cooper instructed me to go up to the fourth floor where I was greeted by Scott and Bradly.

"I saw the news, Scott, why are those people going to the police?"

"I knew there'll be a few," he said, "but not this soon."

"What are you going to do, Scott? Where are you going to take it from here?"

"I'm going to expose you, Sandman. You're going public. You're going to become one of the most famous and well known men in modern history."

"Why me? Why not someone else who's all ready been geneticated-there must be thousands by now?"

"9,456 to be exact."

"Then why me?"

"Because you're unique. You were more . . . how can I say this? Well . . . you were more customizable than all the others. Because you were meticulously cloned, we had more freedom in profiling you. And of course, when we geneticated you, we also had more physiological features of your anatomy that we could exploit. For example, as I have already said, we unleashed more potential in your brain. What we did to you, we couldn't have done to the others. But don't misunderstand, the other natural born people who have been geneticated will possess much of the same attributes as you-but they will never rise to your standard of intelligence and potential."

"How will you expose me to the public, and to be more specific, why?"

"Why? Don't be so pretentious; you know goddamn why. My stealthy little secret is out. It was just a matter of time, but I didn't plan the time would come this early. I anticipated maybe, at the least, another month or more. But so be it-the time has come as surely as I knew it would. And I want you, Christopher, to be our quintessential example of human perfection. You have matured to astuteness and the ability to understand your full potential. You, my dear friend, are not only self-actualized, but you have become a new species . . . and you damn well know it. Thus, you will be our representative because I know you will succeed in convincing the general population that this is not an abomination. After your appearance, the populous of the entire world will beg to be like you."

"So how will you expose me to the public?"

"Through the *World News Network Now,*" Scott said with a shitty grin.

"And will I do this alone, or will there be someone else with me?"

"Dr. Bradly Hues will accompany you."

"What is our assurance of getting in and out of the studio alive? How can you be sure that they wouldn't simply seize us and keep me for further studies?"

"I have your back. I've talked to *Joseph Alans*, president and CEO of the *WNNN*.

"And?"

"He's assured me everything will be on the up-and-up, and if he doesn't comply with my arrangement . . . well, let's just say he'll have to learn a new language."

"What do mean by that?"

"Never mind; just never underestimate my judgment, Sandman. Believe me, you're the safest man in the world. I've talked to Mr. Alans and he's agreed to a bilateral procedure that will present you to the world."

"And what is your unceremonious procedure, if I may ask?"

"You will have a three hour segment that will be prerecorded in a week's advance-seven days from now as a matter of fact. You will demonstrate to an audience and a panel of experts all your capabilities and facets as a new species."

"Could you be more specific?"

"Sure. The audience will be comprised of approximately 400 people who work and teach at various universities. Collectivity, these people will possess the diversity of the human range of experience. As a matter of fact, there will also be one of the most skillful magicians in that panel to make the best case that there will be no trickery."

"And who will this panel of experts be?"

"The finest and best known scientists in the entire world. They will be performing the various experiments on you."

"Give me some examples, Scott?"

"How is your eyesight?" he asked.

"Great; I've never seen with better clarity, and even in the non-visible spectrums of the human eye."

"I'm glad to hear that, Sandman, because one of the experts will be one of the best ophthalmologists' in the world. She will demonstrate a test on your eyesight, which should prove to everyone that you can see beyond human capacity. And how do you feel physically, Christopher?"

"Like *Superman*, in a strange sort of way."

"Good . . . good, and how much weight can you lift?"

"I'd had to say in the neighborhood of about three-quarter ton."

"Splendid," Scott said in a curious tone, "that ought to scare the hell out of them. Do you know what the average bone density is for a human being?"

"Approximately 10 pounds per cubic inch."

"Would you like to know what we've determined yours to be?"

"I know it's more because of the weight I've gained-so what is it?"

"Your bone density has been estimated to grow to about the same value as hematite-do you know what that is?"

"Are you asking me what hematite is or its density?"

"Both."

"Alright, hematite is a naturally occurring ore of iron oxide and its density is roughly 5.3 grams per cubic centimeter. And if I choose to convert that to English units . . . I would approximate my bone density to be 27 pounds per cubic inch."

"Not too shabby; is it? Your bones are just about as strong as iron. Tell me; how's your hearing, Christopher?"

"The last time I measured, it was near 0 to 80,000 hertz in the ultrasonic without regard to intensity. The average human eardrum would burst beyond 20,000 hertz.

"How's your sense of smell, Sandman?"

"It in the range of about 300 to 400 parts per *million*. I can selectively smell your armpits from where I standing if I choose too," I said with a small grin.

"Impressive. What does a dog smell?-About 10 to 20 parts per *billion*?"

"Something like that, Scott."

"Another little thing, Christopher; there will be a test on your immunity system. One of the medical experts will inject you with a lethal poison that kills in seconds, but fear nothing because I know for fact that your immunities will destroy any foreign substance that is not part of your system."

"I can live with that," I said.

"And I've heard that you can hear your body sing to you also-what does that mean, Christopher?"

"The best way that I can describe it is like some type of feedback. Other than that, it's beyond explaining. You should eventually start to feel it also, Scott. You were geneticated too."

"I hope I do, Christopher-I really do," he said introspectively. "Have you been studying?"

"That's about all I do or want to do. I've finished my encyclopedia on mathematics and I'm almost done with my physics and chemistry too."

That's good, because these experts will be testing your knowledge."

"I also have a couple of sound theories on antigravity and propulsion."

"Say that again?" Scott said as he cocked his head quickly to look at me squarely.

"I said . . . I have a couple of sound theories on antigravity and propulsion, and I want the Society to develop the hardware. I've downloaded the plans needed to build these things."

"You're telling me that you've cracked the mystery of antigravity? Do you understand the implications that this secret will bring?"

"Without a doubt."

"This is incredible! If what you're telling me is true, this will help legitimize the Society and bring it into the mainstream of public commerce. We will be transformed from secrecy to legitimacy. This Society has really done nothing wrong, but on the contrary, we have increased the time of mortality and eliminated sickness and disease. It is true that we have cloned, but only from the dead. And we also possess a tremendous amount of science and medicine for our country. Our services would be offered to the entire world and we will prosper enormously in a free market."

"But Scott, what about the man I killed in my guesthouse and the people that the assassins killed? What about the two men that Billy and I tortured in that horror room? And what about all the people you've killed? How can you honestly say that we could possibly become legitimate? And what about all those secret agents?"

"Christopher . . . I don't know what the hell you're talking about."

"Ah, come on Scott; don't act so innocent, you know what I'm taking about."

"Back off Mr. Sandman, and don't forget who you're talking to. You've committed your complete alliance to me, and if you wish to negate your words, then turn around and get the hell out of here. And do it fast."

"Woo! Woo! Scott, please don't misunderstand me. I was merely asking you about some of the underhanded deeds that were done to protect this place. I have absolutely no intention to compromise the Society and all the good things you've done for me. Please, Scott, accept my most sincere apology. I was trying to pose a question, not a betrayal."

"Sorry I jumped the gun, Christopher, but you must realize that I've dedicated my entire life to this establishment, only to make this world a safer and better place. It's always been my intention, when the time was right, to bring to light the fact that we could geneticate a new and better species of humans and convince the population of the validly and importance of this modification. We can heal the sick, the crippled and the lame. We can even eliminate all the infectious diseases including cancer and make the population healthy. Don't you believe that people would want that?"

"I do; but you would also have to tell them that their life span will increase with all the other attributes, including their transition from Homo sapiens to the new species, which you've proudly proclaimed: Homo genesis sapiens-an entirely new race.

"This is true, but I really don't think it's going to be earth-shattering. My guess is that at least 90 percent or more of the population will be begging for the procedure. And what's not covered on their insurance, I'll do for free."

"How are you so sure that the government won't come down hot and heavy and put you and this whole goddamn place away?"

"We have a little over two hundred certified and licensed experts in all the fields of specialized human and animal medicine, anatomy, genetics, biochemistry, and all the way down to our attorneys, computer gurus, mathematicians, physicists, material scientists, construction workers and janitorial services-and that doesn't fill in all the positions in between. We are, in all considerations, the most legal goddamn fucking set of professionals put together under one structure. We are as legal as legal can be. At this very moment, we have a team of attorneys working together with the state's legislative body to pass a bill and permit us to continue our genetication to anybody who wishes. The bill has already passed through the Legislative Reference Bureau and the Chief Clerk has assigned the proposal as House Bill 648. Presently, the House Speaker is assigning the Bill to a Standing Committee, and will be made to the public when we air you on the *World News Network Now*. By then, Bill 648 will be in the process of the three-day consideration by the House Floor. After that it goes to the Senate, which I'm confident will pass without amendments, and then I'm estimating that the Governor will sign it into law before year's end-we will then be an officially permitted business."

"You seem very self-assured, Scott."

"Indeed, I do. I have many friends that you don't even know about. And I have friends in many governments, or have you forgotten?"

"No, Scott, I haven't."

"Well then, enough of this talk, Christopher, let's put it behind us. I'm looking forward to have our physicists read your files and plans about your propulsion and antigravity theories. I have a very strong inclination that they'll work. Thanks for you co-operation, and I'll see you in a week."

I didn't get home until around 10:30pm and when I walked through the door, the first thing I saw was my wife sitting in a chair reading.

"What are you reading, sweetheart?"

"Something that I found in a box that I packed, it's your old manuscript: *The Calabi-Yau Universe*. I just finished it and I must say, I thought it was pretty good. I forgot all about it-sorry I didn't read it sooner, but like I just said, I forgot.

"Did you like the ending?"

"That was the best part-I believe it's publishable, why haven't you pursued it?"

"I have better things to write than science fiction."

"You're talking about the Society; aren't you?"

"Yes."

"Speaking of the Society; what did Scott want?"

"I'm going to be on *World News Network Now* next week."

"You're kidding-what for?"

"To expose myself and the genetication process to the world."

"Why would he, or you, want to do this? I don't understand?"

"To unveil his personal hidden agenda and show the masses how much better life would be if he could enhance and change the population through genetication. He appears determined to make it happen. I could call him a madman or a genius, but I believe his intentions are necessary if the world is to survive its growing despondency. I view Scott as a futurist and a genius-and everything I've learned and discovered about him, proves to me he's right."

"When will you appear on *WNNN*?"

"In about a week. But it'll be prerecorded and will air the following week."

"Christopher, honey, are you sure this would be a good thing?"

"Absolutely. The year has come to move this world forward and to the future. Genetication is real; it's happening, and it's good. It will heal the weak and give prosperity to the poor. It will give the general population the greater intelligence it needs. It will give them the capable brainpower to improve their deplorable environments and conditions and become involved and concerned in their own success and wealth. It means jobs that are not just inconsequential, but jobs that require skill and knowledge. Genetication will raise the standard of living throughout the world. Can't you see the truth in this, sweetheart?"

"When you say it with such determination, I suppose I can."

"Let me ask you a few questions?" I implored Darlene.

"I'm listening, honey."

"How have you been feeling lately? How are you senses? How's your hearing, your eyesight, your strength and other things?"

"I have to agree with you, Christopher. I feel very healthy. Everything has become more acute, and as far as my intelligence, I also have to agree-I understand and comprehend much more then before."

"So, would you say that being geneticated was a good thing?"

"Absolutely."

Darlene and I spent the next six days doing whatever young lovers probably do. I spent some of my afternoons working on completing my physics and chemistry encyclopedias while I saved the mornings and evenings for Darlene. She continued with her painting and painted the most beautiful scenes of trees and nature that I've ever seen. Most were so extraordinarily surreal that just looking at them memorized me and put me in a state of peace that I couldn't believe. She had two matted and framed, and displayed in our living room adjacent to our fireplace. There were a couple of afternoons that she went out to do some shopping for herself and for me. Most of the items she brought back were exquisite and worthy possessions that helped decorate the place. She was starting to make a house our home.

It was five o'clock on a Sunday morning when I received a call that I had been anxiously awaiting, but yet dreading.

"Hello, Scott," I said in a waking voice, "what time do you need me?"

"I'll have a limo there by 6:30am. Today's the day that we'll make history. Be on your best and I'll see you soon."

"Was that Scott?" Darlene asked as she turned over.

"Yes, that was Scott, sweetheart-close you eyes and sleep; it's early."

"I want to get up with you so you won't be alone. I'll make us some coffee."

"If you wish, sweetheart-thank you. I'm going to shower and dress-I'll be right down with you."

When I walked through the kitchen and into the dining room, I noticed that Darlene had put on her silky morning grown and was patiently waiting for me as she sipped on her coffee while flipping the pages of a magazine. She looked up as I walked in.

"What time are you leaving?" she asked softly.

"Scott will be here at 6:30am to pick me up."

"Will you be gone for more then a day?"

"That's hard to say, sweetheart, I don't know what the agenda is. I presume that I'll be back sometime tonight. If not, I'll definitely call you. Sweetheart, I sense sadness-what's wrong?"

"I'm just a little worried for you, and for me. I feel that this is the end of something, and yet, strangely, the beginning that may change us somehow forever. I fear something, Christopher."

"Fear nothing my dear, dear, sweet Darlene," I told her as I bent down and kissed her cheek. "We will always be together. I love you more than yesterday, and I'll love you even more tomorrow."

"There's a car pulling up the driveway," she said looking at a monitor.

"That's my ride, sweetheart. If I'm not coming home tonight, I'll call you. Love you babe."

"I love you too, honey-please be careful."

I couldn't quite put it into my thoughts, but I felt a sorrow for my wife. It felt as if I should turn back and forget the whole thing-to simply quit and walk away. But I knew that that would be impossible, I had an important job to do today. The limousine door opened, and in the back seat sat Bradly.

"Welcome to the day that will change the days for years to come," he said. I climbed in slowly and sat across from him. His appearance radiated majestically and then his words echoed with buoyancy. "Scott's very proud of you today, Christopher, because today you will validate everything he's worked for. He has the utmost confidence in you-he'll be watching on closed-circuit television."

"How long will it take to get this circus underway?" I asked wishing the whole thing would go away.

"Oh, about twenty minutes to the airport, and then a forty minute flight to the studio," Bradly said, and then smiled.

"Scott thinks I'm going to change to world. Do you think I can do that in three hours?"

"Christopher, not only will you convince these experts and the audience within those three hours, but by the time this airs, you're going to turn the world upside down; backward and forward, left and right and inside out."

"That's what I'm apprehensive of," I said, knowing it was all too true.

"You sound like this is a bad thing. Why do you portray it in that sort of way?"

"Up to this point, Bradly . . . Scott and all his cohorts have been doing astonishing and incredible research, development and constructing amazing unearthly-like discoveries. The Society is an enigma that's far ahead of its time, and how it has remained a secret still mystifies me. But I'll tell you right now, and mark my words, after this show airs, it'll be on the tongues of people all over the world. But it won't stop there; no sir indeed.

Mass discord and civil disobedience will ripple the surface of this earth. The ramifications that I foresee may be formidable. Percentage wise; I give it about a 50/50 chance that this earth survives or dies.—Is that the airport over there?"

Bradly looked at my pointing finger, still pondering my words momentarily with perplexity and uncertainty, and then said, "Yes, we're here."

The limo made a turn and drove through an open gated enclosure and then headed down a tarmac where an unmarked black jet was parked. It wasn't a big jet, but neither was it small. Its engines were throttled vibrantly but yet not strong enough to go. I followed Bradly up the steps and as I walked inside, the doors behind me closed. Bradly took a seat, buckled up, and then pointed one for me. Slowly the jet began to move and as I looked out the window beside me, I could see it was turning to straighten itself along a line. Suddenly the engines screamed and the thrust pushed me in my seat. I felt the little wiggles and bumps as it began racing down the runway until very quickly a smooth inclined ride was all I could feel.

"First time on a jet?" Bradly said mannerly.

"Yes, and it's very enthralling-I like it."

"Good for you," he replied—in the future you'll enjoy many more rides. Some more fantastic than these.

In less then forty-five minutes of flying enjoyment, the land outside the window tilted the other way as the jet was preparing to land in a city I've never been before. We taxied around and rolled to a stop where another limo was standing by. And like studious businessmen, we exited the jet and climbed in a limo that drove us to a final stop. It was in the heart of the city and outside of a tall building where a marquee that electronically kept scrolling the words: *World News Network Now*. When we stepped out, there was a man to greet us and introduced himself as, Mr. Joseph Alans.

"It's a pleasure to meet you Dr. Hues," Joseph said as he shook Bradly's hand.

"Please, Mr. Alans, just call me, Bradly."

"And you may simply call me, Joe," he replied, and then turned to me and said, "And may I presuppose that this is your special man, Mr. Christopher Lance?" he said with a smile and an extended hand.

I shook his hand and said in kind, "Please, Joe, just call me, Christopher."

"It's an honor to meet you, Christopher. Shall we go inside to the studio-the show is scheduled to go in less then an hour."

We sat in Joe's office and signed some papers as he told us what to expect when the show began. He reiterated that the audience will be comprised of approximately 400 people who work and teach at various universities and that the panel of adjudicators and experts will contain 20 people. Joe also told us that the panel experts would be performing the various experiments on me and that I wouldn't know what to expect as the show went on. Finally, it was time to present myself on stage and eventually, to the world. Bradly sat in the wing as I followed Joe through a curtain and onto the stage. Joe began to speak.

"Ladies and gentlemen-today, on WNNN, I want to present a phenomenon that's standing right beside me. His name is Christopher Lance, and he's here today to show us a glimpse of a possible new future. Christopher is the first ever-cloned human. And as if that isn't remarkable enough, let me tell you folk's one more additional detail-he's been geneticated to give him some incredible characteristics.

Today, as you can see, we have a panel of expects from all over the world who will be performing some extraordinary analysis on his physical, mental, and sensory features. And know, without farther ado, I present to you this incredible man, Mr. Christopher Lance."

After a short applause, one of the experts approached me and said, "Please present the palm of a hand and hold it right-side up and toward me."

I complied with his demand as he then set a stone in my hand and told me to crush it. The cameras zoomed in to watch. I looked at it for a moment and realized it was nothing more than the sedimentary, grayish-white rock called, periclase. Although an ordinary human hand couldn't crush it, I knew that mine could. I crushed it and let the dust and pieces fall to the floor. I then heard the mumbling from the audience, and the chatter amongst the panel of experts. The man who placed the rock in my hand looked at me bug-eyed, and then approached the panel with his arms and hands bent in the air trying to utter words where none could be found.

As he sat down, another expert came forward and unveiled a set of barbells. They varied in weights from 500 lbs to 1000 lbs. "What I'd like for you to do, Mr. Lance, and in a final standings position, lift that 500 lb weight above you head and hold it there for five minutes."

I did as he said without a bead of sweat. "That was quiet a remarkable determination of strength," he said, and then continued, "now do the same with the heavier weight." I repeated my performance with just a slight bit more effort, but I had fulfilled what he had asked. Holding half of ton above my head for five minutes convinced the panel of experts something that an ordinary man could never have done. He sat back down with the others shaking his head with wonder and disbelief.

After a few moments, a lady expert rose and walked toward me. "Please. Mr. Lance, have a seat right here behind the sign above you." And then she exposed in front of me, a flat plate of microscopic optic fibers closely adjacent and touching. She flipped a few switches; turn a knob that lit the optic plate to show the spectrum of visible wavelengths. She turned around and said, "Does everyone see the rainbow of colors?" There was a murmur of "yes" from the audience and experts. "And what does the sign display read which Mr. Lance can't see."

"4000 to 7000 angstroms," the audience and experts rang out almost in unison.

"Now we're going to play a little game with colors," she said loudly for all to hear, and then turned her face to the crowd and said, "Please, just for the audience, from this point on just watch the sign display, but do not yell out any answers. She then turned to me and said, "Tell me and the audience if or when you see any colors." Suddenly, the optic plate turned clear. The game was about to begin.

"Mr. Lance, do you see any colors?"

"No."

"Alright, Mr. Lance, do you see any colors now?"

"Yes," I said, "I see a violet color which I might estimate at 3200 angstroms."

"Okay, Mr. Lance, do you see any colors now?" she asked.

"I see two. Violet at about 2950 angstroms on one side, and red at about 8420 angstroms on the other."

"And what about now? What colors do you see?"

"Nothing," was my response.

"And now?" she asked.

"I see only one, but with very little color; about 75 angstroms. I'd say you're showing me some color near the x-ray spectrum."

"Do you see any color now, Mr. Lance?"

"A very light red; infrared. I calculate approximately 9500 angstroms."

Suddenly, she blurted out, "This experiment is over. I do not understand how you did that, Mr. Lance. Not only did you see the colors, but you also managed to actually call out their wavelengths. To me, this is far beyond human capacity. I proclaim that something is very momentous about you, sir, but it is far beyond my grasp."

The sign display that the audience saw apparently agreed with me, and after the lady sat down, the crowd began to applaud and cheer. After a few moments, the noise slowly grew to silence again.

Another expert from the panel got up and approached me with another test-an olfaction experiment, which compared my sense of smell with that of an olfactometer. In all cases, I was quicker at smelling different odors at a given distance than the meter. Again, the expert walked away dumbfounded.

Then came a real challenge. Two experts from the panel came forward and retrieved a cow from behind the curtain. One of the experts drew two hypodermic needles partially full with six millimeters of a deadly solution of poison from the same container. "When injected," an expert said, "death begins very soon, but unforgivably, it is also extremely painful and bloody."

The expert injected one of the hypodermic needles into the cow's neck, and then the other needle in mine.

Within five seconds, the cow fell down and starting kicking very fast and hard. The poor thing had blood running from its eyes, nose, mouth, ears, and even from the thin-skinned fleshy parts of its body. It died a horrible, gruesome and bloody death. After the cow was dead, the crowd gasped with utter horror mixed with astonishment and contempt.

Then the poison was injected into my neck, but it had no effect on me. There was an astonished mumble from the crowd.

The last thing they did was to quiz me on some of the most difficult, and diverse questions that the experts, and the audience could think of. But most were in the realm of science. Some of the questions I could simply answer with words, while others I had to answer with equations, symbols and diagrams that I displayed on a large plasma screen.

After all was through and done, I was given a standing ovation and then a rather fast escort off the stage and behind the curtain by the show's host: Mr. Joseph Alans. In one and a half hours, Bradly and I were standing inside the Society, and facing Scott.

"You were absolutely magnificent, Christopher! Next week, when this show airs, you're going to be headline news. I can image that the people who watch the show will want to be like you. You are the solution to the world's problems. I'm very proud of you, Christopher. You did a very good job today . . . very well done indeed.

I'm sure you're tired and would like to leave now-it's certainly been a long day for you. I'll have a driver take you home so you can get some sleep and I'll stay in touch."

Scott was right. When I got home, all I wanted to do was sleep. However, I stayed up a few more hours to talk to Darlene. I told her everything, including my concerns about humanity. She took everything very seriously, which kind of troubled me. We talked a while more and then I told her I had to sleep.

"Please wake me up when you wake up, I've got work to do," I asked her nicely. She kissed me goodnight as I did with her. "I promise to spend some time with you too, sweetheart."

Darlene awoke the next morning after I had about fifteen hours of sleep.

"Wake up, my sleepy head, you asked me to wake you when I woke up," I heard her voice say in a loving tone.

"Good morning, my baby," I said to her as I rolled over.

"And good morning to you, honey-how are you feeling?"

"Very rested, thank you. Am I too nasty to roll you onto me?"

"You're never too nasty for me," she said as she rolled me onto her.

We made love for a couple of hours as we whispered immeasurably our passions for each other.

To me, as I told to her, she was my true goddess, as she had declared to me, her love would be forever. It was such a beautiful morning to us as we unionized together and as we stroked each other in the most pleasant and softly ways. We were very deeply in love with each, as she felt and which I knew, that no humans could ever feel. She told me her body was singing to her, as mine was singing to me too.

In privacy and in respect, and after our necessary bodily needs were through, we took a shower together and washed each other in naked solitude. Darlene was in ecstasy as she washed my body, and I was truly in awe, as I washed her body too. It was crystal-clear to me, what this genetication could do; and now, when it came to love, I was entirely surprised how much it would do. It was such an extreme feeling that I honestly knew that humans truthfully could never feel.

CHAPTER TWENTY TWO

"How's your coffee?" Darlene politely asked me.

"Just the way I like it, but not as much as you," I said as I patted her butt.

"You told me you were going to work today; what are you going to do?"

"I'm just going to download my finished encyclopedias on chemistry and physics."

"You amaze me, Christopher. You've managed to put together three complete sets of encyclopedias-what are you going to do now?"

"I want to ascertain a new science of exploration of space. I call it: *spacetime-deformation and massless acceleration.* It adds more to the principles of general relativity due to a new phenomenon, which I discovered in antigravity. I've downloaded the necessary documents to the Society, but there's more that I need to see. I want to build a starship that bends reality into one dimensional space."

"I know that I understand much more than I use too, and I can read books in a matter of hours, instead of days, but I don't think I'll ever comprehend the amount of material you treasure and appreciate."

"Are you going to do any painting today? Paint me a picture of a glimmering white photon that splits into a rainbow of light."

"Would you like me to paint that for you?"

"I'd love to see you try."

"Then that's what I shall do today."

For the next seven days, Darlene and I lived a life of isolation and love making, taking walks and talking silly small talks, while sometimes dressing up with occasional shopping sprees, picnics in our garden and fine dinning. We were in love and having the time of our lives, until on that seventh day, the phone rang.

"Hello . . . Christopher speaking?"

"Who were you expecting, Christopher, the sandman? Did you see the show on WNNN?

"No . . . I was there if you'll recall, why would I want . . ."

"I'd like to see you within the hour. This is a serious matter and I want to talk," and then all I heard was a click on the other end.

From the sound of Scott's voice, I knew he meant business. This time, I didn't drive Darlene's car, but instead, drove my Jaguar. I paid little attention to stop lights or signs, and avoided all the other cars until I made it to the mountainous roads. Then I gave my Jag hell, leaving nothing but rubber and smoke and a trail of black tracks. In less then an hour, and without any traffic events, I was at the Society and facing Scott. "I busted my ass getting here, Scott; I really did. You have my full attention, sir."

"Christopher, I wished you had watched the show. Before it was over, there were at least a thousand or more people waiting outside that building just to get a glimpse of you-they didn't know that the show was pre-recorded. And when the show was airing, WNNN was running a live poll to see who would want this procedure done to them. Would you like to know the answer?" he said kindly.

"Sure, tell me."

"Ninety-eight percent wanted what was done to you, done to them. Does that surprise you?"

"Initially, no. But I don't think that most people would consider the long-range consequences of their actions-how would societies react? Have you thought of that, Scott?"

"I have."

"And?"

"Well, let me think . . . okay, for the good, and for the bad, I first would expect some kind of economic upheavals and a division among the population between the rich and poor. Racism might start to run amok and religious congregations may split and separate into radical fractions of pros and cons with some blaming the Society as an abomination of evil and wicked people plotting to destroy the world. There would be fears of over population, especially when genetication spreads to other countries. Where would we find the food to feed all the people?

Then come the jobs. Some would view them as selective and unfair. Think about law enforcement-who would want an average human when a geneticated human would do much better? And what about manual labor and heavy construction work; who would do the better jobs? Sports and other athletics would desire the new geneticated humans. Corporations would look for the most intelligent people; pharmaceutical companies might even become obsolete. Medical doctors would be on the downswing, as well as insurance companies, funeral homes and cemeteries.

Universities would have to raise their standards, without prejurious, because of the rise in the intelligent quotients amongst the population, and new advances in technologies would be their progression. Even the United States Constitution would most likely change. Maybe the juridical systems could become backed-up or over crowded, not to mention the questionability of prison sentences and the inferiority of the non-geneticated human being. There's a lot to be considered, Christopher-wouldn't you agree?"

"I do indeed, but what would the best time in a human's life to be geneticated?"

"I've concluded, and Karen has agreed, that starting at the age of one to no more then the age of sixty. Of course, there will be exceptions. And we will no longer be considered human beings, but rather, genus beings. Although I know that there will be a considerable amount of disparity, I'm positive that our new race will eventually emerge."

"I know what you're telling me, Scott, but what was your real meaning for bringing me here?"

"Follow me to see your new machine," he said respectfully.

We entered a room deep inside the mountain and there floated a shiny metallic machine that looked like a flatten ellipsoid. "Your physicists and engineers have succeeded with my antigravity theories."

"This is just a prototype; we're going to need your help in developing a larger and stronger device. This is what I wanted to really show you, Christopher, your theories have made this thing a reality, but we need you to teach the scientists and engineers about the mathematics and the physics which you have evolved."

TWENTY YEARS LATER . . .

Although I had never seen him again, Scott had it right all along. There was turmoil and strife in the genetication of the world. There were three consecutive wars that took its toll on planet earth, but after a twenty-year span there was indeed, a rebirth. Darlene and I had two children: Alyssa and Scotty. Both are very smart and collectively superior than we are, and although twenty years have gone by, they are still at the age of six and five, respectively.

I've seen the people who were not, or refused to be geneticated grow much older then me, and some have even died. But they're still around and will be for about another fifty years, even though they can't compete with the new race of genus beings. As the world continues to transform, there's also a noticeable amount of less terrorism as we all begin to grow wiser with more and more geneticated sages.

Through the years, many things have changed-especially for the genus beings thirteen years or older-deemed the new legal age. But there are still some human beings, and some genus beings that still kill others, statistically, about one out of every thousand. Yesterday, a genus was sentenced to exchange her life for those whom she'd killed.

She was sentenced to die as provided under a new amendment sanctioned by Article V of the constitution. She was executed by way of expulsion from the genus race, as a matter of fact, even from earth. And in doing so, she became an integral and new source of life-she became energy. Her organic mass was converted to pure energy, a system much more efficient than combustion-and it generates over 30 trillion watts of surplus power each year. It appears to be the common consensus of the people-send the calculating killers to hell. And as far as I'm concerned, having your body desecrated into a plasma of electromagnetic radiation just didn't sound like heaven to me.

Darlene and I still live in our beautiful home, while Scott and Bradly still remain at the Society, which continues to still be a mystery to many. Genetication is now carried out in many separate complexes scattered about the world.

Billy is still very active with his chain of stores, which has made us much richer then twenty years ago. He remarried and still remains one of my best friends.

There are no cars that roll on wheels and burn ethanol anymore. They have been replaced with my flatten-ellipsoid, flying micro-spacetime-deformation vessels, commonly referred to as, *Wolphoes,* because of the soft noise they make when they fly through air. Besides the Mars and the spacebase communities, these interesting vessels became the backbone of the world's space programs.

The United States of the World was just beginning a project to build a starship based on this technology. From the ground at night, the starship is easily visible with the naked eye, except for the eyesight of a human being.

It was being constructed close to mars, and it was becoming one of the foremost goals that genus beings had in the plans, and I can proudly say that I initiated the start of this dream.

It was a beautiful summer day, while working in my office that the teledisplay rang. I pushed a button and Bradly's face appeared.

"Afternoon, Bradly, what can I do for you?"

"Christopher? How much do you know about crop circles?"

"They were very common near the turn of the century, but most, if not all, were proven to be hoaxes. They're very rare today . . . why?"

"Take a look at this one," Bradly said as he showed me a 3D view of a very strange formation in the Pacific Ocean. It was taken by the world's largest landstat satellite and it occupies more than one thousand miles in diameter and standing waves of 100 feet high.

"Boy, that's a big one . . . the biggest and most complex I've ever seen."

There's something very strange about this one Sandman," Bradly said. "It appears to have no pattern or discernable designs, and we have satellite conformation that it was made in the light of day and in the span of less then fifteen minutes. We would like you to study it and then tell us what you think?"

"Very well, Bradly, download it and I'll take a look."

It was indeed, a very large and complex display of unusual water standing waves-never before seen in the archives of history. The more I studied it, the more intrigued I became. After hours of looking at it in many different views, it suddenly popped out in front of me. It was written in Sumerian, Latin, Hebrew, German and binary digits. After I deciphered it, I called and displayed it to Bradly.

"Good evening, Sandman; any news? What can you tell me about that strange figure and does it have any significance to our planet or people?"

"Yes. It's alien. It was made from a race of another world."

"What do you mean? Explain."

"The *first part* was designed in the Sumerian language, which when translated said: "Welcome, people of earth. And in Latin it said, "The pulse counts down in fifty years. In Hebrew it was written, "We bring new offerings from Scott Sellers". With the binary numbers, it presents itself as celestial coordinates designated as right ascension of: 6 hours 25.321 minutes and a declination of -8.141 degrees. I would suggest that someone point a telescope in that direction.

Later that evening, Bradly displayed to me and told me that my conclusions were confirmed to be 99.99 percent correct.

"It's pulsating down, Christopher. It's a recent star-like object that's never been mapped. According to our astronomers, it's pulsating down like a clock. We've timed it with our atomic clocks and it matches to within 99.99 percent of earth's time. And if we calculate backward to the first time we observed that water crop circle, we find that it's counting down to zero."

"So what do we do, Christopher—how should we respond?"

"We needn't have to, but instead, we'll reschedule our plans to complete our Starship and meet our Alien Originator, Scott Sellers, halfway here."

END

AUTHOR CREDITS

Lansing Community College, Lansing, Michigan, *Associate of Applied Science-Drafting*, 1973; Minor: Civil Engineering.

C.S. Mott Community College, Flint, Michigan, Associate of Science-Chemistry, 1979; Minor: Physics and Mathematics. Douglas Graduated with Honors.

University of Michigan, Flint, Michigan, Completed advanced Mathematical courses in Quantum Physics of Atoms, Molecules, Solids, Nuclei, and Particles. Also completed advanced Mathematical courses in Concepts of Modern Physics, i.e., Atomic Structure, Radioactivity, Elementary Particles, Relativity, Statistical Mechanics, and much more.

Douglas has also consulted many institutions and professionals in many areas of Computer Aided Drafting; including General Motors, The City of Rochester Hills in Michigan, Manufactures, OEM's and many others.

Douglas is also a member of the **FineArtAmerica.com** community (FAA). He has on his exhibit over 25 watercolors to display and sell.